The Challenge of the Shroud
History, Science and the Shroud of Turin

Mark Oxley

authorHOUSE®

AuthorHouse™ UK Ltd.
500 Avebury Boulevard
Central Milton Keynes, MK9 2BE
www.authorhouse.co.uk
Phone: 08001974150

© 2010 Mark Oxley. All rights reserved.

No part of this book may be reproduced, stored in a retrieval system, or transmitted by any means without the written permission of the author.

First published by AuthorHouse   4/1/2010

ISBN: 978-1-4520-0009-1 (sc)

This book is printed on acid-free paper.

ACKNOWLEDGEMENTS

It is not possible to write a book such as this without assistance and support from many people. I have been fortunate in the kindness I have received from friends in Harare and from authorities and researchers on the Shroud around the world.

Ian Wilson is undoubtedly the doyen today of Shroud historians. His book, *The Turin Shroud,* published in 1978 broke new ground in investigating the history of the Shroud. His later book, *The Blood and the Shroud,* rekindled my interest in the subject and was the direct inspiration of my decision to set about writing this book. Ian has been unfailingly courteous and generous in responding to my many queries about the Shroud and in encouraging me in this project.

No serious study of the Shroud can be made without reference to Barrie Schwortz and his website, *www.shroud.com.* This is an invaluable resource to anybody who wants to keep abreast of historical and scientific research on the Shroud. Barrie, too, has been unfailingly generous in his response to my many queries and in granting me permission to use his photographs from the Shroud of Turin Research Project in 1978.

Marc Guscin is another scholar and researcher whose assistance to me has been invaluable. He was kind enough to send me a copy of an unpublished manuscript of a book that he has written on the subject of the Sudarium of Oviedo, for which I am most grateful.

There are numerous other scholars and researchers who have replied to my enquiries promptly and generously. Prof Dan Scavone, Prof Karl-Heinz Dietz, Cesar Barta, Alessandro Piana, Joe Marino, Diana Fulbright, Dr Gilbert Lavoie and Rev Ken Stevenson have all responded to my enquiries without hesitation and have all been supportive of my work.

Special mention must be made of the assistance I have received from Dorothy Crispino, who has sent me both old issues of her newsletter Shroud Spectrum International at my request and several unpublished documents that I found invaluable in trying to unravel some of the mediaeval history of the Shroud.

There are other writers on the Shroud with whom I have not been in contact but whose work has been of great value and interest to me. These include Prof Bill Meacham, Mark Antonacci and the late Noel Currer-Briggs.

I would particularly like to acknowledge a debt to the late Ray Rogers, whose work in the field of chemistry has made a major contribution towards a better scientific understanding of the image on the Shroud. Although I was never in contact with him I have been inspired by his work and his dedication to the search for scientific truth.

Closer to home I would like to acknowledge the support and friendship I have received from Fried Lutz and from Derek Huggins and Helen Lieros at Gallery Delta in Harare. Derek introduced me to the mysticism of Orthodox Christianity and Helen continues to inspire all who know her with her artistic talent.

Fr Steven Buckland SJ, the former Dean of Arrupe College, the Jesuit School of Philosophy and the Humanities in Harare, kindly allowed me unfettered access to the Library of Arrupe College, which was of great assistance in my research into early Church history and historical events surrounding the Shroud.

I would particularly like to express my thanks and appreciation to Rosie Mitchell for the superb maps of the Middle East, France and Italy she produced to illustrate where so many of the historical events described in this book and associated with the history of he Shroud took place.

This book is divided into three major sections – one on the Shroud and History, one on the Shroud and the Crucifixion and one on the Shroud and Science. The section on the Crucifixion is difficult and graphic but I believe that an understanding of crucifixion, from both a medical viewpoint and the viewpoint of the Gospel writers, is essential to an understanding of the image on the Shroud. I must acknowledge my debt to the research and writings on the subject of crucifixion of Dr Pierre Barbet and, more recently, Dr Frederick Zugibe. I have drawn heavily on their work.

Where I have referred to or quoted the work of other writers, I have endeavoured to represent their views accurately. If I have failed to do so the blame is entirely mine. The conclusions I have reached based on the research I have done are entirely my own and I take full responsibility for them.

In concluding these acknowledgements I would like to quote two observations that have been made to me, the first by Marc Guscin and the second by Prof Dan Scavone.

The Shroud does not and cannot prove the Resurrection in any way. It is a piece of cloth with an enigmatic image on it that may or may not provide evidence of an event that took place some 2 000 years ago. It is up to each and every interested individual to reach his or her own conclusions about the meaning and nature of the Shroud – this is the challenge in the title of this book.

History proceeds from documents, not arguments from silence. Historical research must be based on documents. Without documents one can only surmise and hypothesise without reaching any valid conclusion.

Mark Oxley

Harare, Zimbabwe

October 2008

Map 1

Map 2

Map 3

Contents

INTRODUCTION ix

PART 1 – INTRODUCTION TO THE SHROUD
Chapter 1 – The Shroud of Turin 3

PART 2 – THE SHROUD AND HISTORY
Chapter 2 – Jerusalem, Antioch and Edessa 13
Chapter 3 – From Edessa to Constantinople 28
Chapter 4 – Geoffrey de Charny and the Shroud of Lirey 44
Chapter 5 – The Later Years in Lirey 54
Chapter 6 – From the de Charny Family to the House of Savoy 64
Chapter 7 – To the Present Day 73
Chapter 8 – The Missing Years: Cathars or Templars? 86
Chapter 9 – The Missing Years: From Athens to France 100

PART 3 – THE SHROUD AND THE CRUCIFIXION
Chapter 10 – Death on the Cross 119
Chapter 11 – The Passion and Death of Jesus Christ 157
Chapter 12 – The Evidence of the Man on the Shroud 169
Chapter 13 – The Sudarium of Oviedo 182

PART 4 – THE SHROUD AND SCIENCE
Chapter 14 – The First Investigations 195
Chapter 15 – The Shroud of Turin Research Project (STURP) 207
Chapter 16 – Dating the Shroud 221
Chapter 17 – Image Formation: Radiation Hypotheses 237
Chapter 18 – Image Formation: Contact and Chemical Hypotheses 252
Chapter 19 – Preservation, Restoration and the Future for Research 261

PART 5 – THE CHALLENGE OF THE SHROUD

Chapter 20 – Heroes, Sceptics and Challenges   277
NOTES   283

Introduction

I first became interested in the Shroud of Turin as a schoolboy in the 1960s. I attended a Catholic school, St George's College, in what was then Salisbury, Rhodesia. I was a boarder. From time to time films were shown in the school hall for our entertainment and sometimes our education. One night the film was about the Shroud of Turin. It featured the World War II bomber pilot Group Captain Leonard Cheshire explaining the Shroud, its history and what at that time was known about the markings on it.

As a schoolboy I was a voracious reader. I had a particular interest in books about World War II, because both of my parents had served in the British Army during that war and they had told me quite a lot about their experiences. I had read Group Captain Cheshire's wartime memoirs, so I knew that he was a hero and a holder of the Victoria Cross. This ensured for me that anything he said obviously had to be reliable.

I was fascinated by what he had to say about the Shroud in the film we saw. He explained how it was effectively a photographic negative and that, when a photograph was taken of the Shroud, the negative of that photograph produced a positive image of what seemed to be a scourged and crucified man. He explained how the Shroud had a history going back to the fourteenth century yet exhibited phenomena that were not known until the nineteenth or twentieth centuries. These included the correct flow of blood. This indicated that it could not be a fake, as how would any forger of the early Middle Ages know how to depict accurate blood flows? He also explained some of the physical aspects of crucifixion – which fascinated a group of schoolboys like me – and particularly referred to the way in which the thumb of a crucified person would close over his palm as a nail was driven through his wrist and touched a particular nerve in the wrist. He showed how this was the case with the man on the Shroud.

Mark Oxley

I have no doubt he mentioned other aspects of the Shroud, but these particularly remain in my mind. As a Catholic boy and an admirer of Group Captain Cheshire I had every reason to believe that the Shroud was indeed the burial cloth of the crucified Jesus Christ. To be quite open about it, and to declare an interest at this early stage, I have never had any reason to change my mind.

Quite by chance, some years later I had the opportunity to meet Lord Cheshire, as he had then become, when he was visiting Cheshire Homes in Salisbury. I told him about having seen his film and how it had had an impact on me. I think he was gratified.

My next encounter with the Shroud came, in all places, in the Readers' Digest. Sometime in the early 1980s – I do not remember the exact date – I read an issue of the Readers' Digest that had as its Book Feature a report on the Shroud of Turin Research Project (STURP). By this time I had obtained a degree in science from Trinity College, Dublin and the scientific details of the study fascinated me. They bore out everything that Group Captain Cheshire had said in his film years before. Clearly the scientific evidence was also pointing strongly to the conclusion that the Shroud was the authentic burial cloth of a man tortured and crucified in the first century in the manner described in the Gospels. Of course, what was still missing was the final proof in the form of a radiocarbon dating test.

My science degree included physics and chemistry, so I was quite familiar not only with the techniques described by the STURP team but also with the theory and significance of radiocarbon dating.

Two points from the report in the Readers' Digest have remained with me over the years. Firstly, the result of three-dimensional images taken by the team showed that the image on the Shroud could not have been formed two-dimensionally, as a painting would be. In other words, the Shroud could not be a painting. Secondly, a three-dimensional image of the face of the man on the Shroud looked remarkably like all the representations I had ever seen of Jesus Christ. This had to be more than coincidence.

What had also occurred to me by this time was that the Shroud must be an extremely disturbing and challenging item for non-Christians, including the many agnostics of Europe. If the Shroud truly was the burial cloth of Jesus it would seem to be evidence that he did die and rise again, as the Gospels claim. In which case he might even be the Son of God, as Christians believe. This would put a hefty dent in the comfort zones of modern agnostics and would be a direct challenge particularly to those who see Christ as no more than a vague historical religious teacher. It would not be at all surprising if many people tried to discredit it or even physically destroy it.

In 1988 three pieces of cloth taken from the Shroud were subjected to radiocarbon testing at three separate laboratories. The results of all three tests, announced in October that year, indicated that the Shroud's raw flax had been made into linen about the year 1325, plus or minus 65 years. In other words the material of the Shroud itself was no more than 700 years old. This was greeted by the international press and everyone else who had expressed scepticism about the Shroud as "proof" that it is a mediaeval forgery.

These tests were carried out under rigorous scientific conditions and I certainly had no reason to believe that they were faked in any way. From everything I had read the scientists and laboratories involved were of impeccable reputation. But I had to ask myself – if 95% of the scientific evidence, in the form of the STURP results, point in one direction, and 5%, in the form of the radiocarbon dating results, point in the other, does one challenge the 95% or the 5%? I could reach no other conclusion than that there must be something wrong with the radiocarbon dating and that time would in due course tell what it was.

Again years passed. I had read books on the Shroud and related topics, such as other items believed to be relics of the Passion of Christ. A few years ago I obtained a copy of Ian Wilson's book *The Blood and the Shroud*. This mentioned the theory of an American microbiologist, Dr Leoncio Garza-Valdes, announced in 1994, that a bioplastic coating on the Shroud material could be the cause of a major error in the radiocarbon dating[1]. This was part of a greater scientific debate on the dating of the Shroud which has still not reached any final conclusion and which this book will attempt to describe in some detail in Chapter 16. Ian Wilson's book in general and Dr Garza-Valdes' theory in particular stirred my interest once again, leading me to want to find out yet more about the Shroud, its history and the scientific studies that have been carried out. Eventually they led to my decision to write this book.

The Shroud presents many challenges. It challenges those who claim it is a mediaeval forgery to replicate it. Nobody has yet been able to do so with any credibility. This must be an argument in favour of its authenticity. Could a fourteenth century forger, with the limited scientific knowledge of his time, really produce an artifact that can still not be replicated by all the wonders of twenty-first century science?

In particular it challenges scientists. The Shroud is said to be the most studied single artifact in history, but still scientists cannot give an indisputable answer to the question of how the image on the Shroud was formed. Scientific papers on the Shroud are produced, and challenged, every year. The Shroud even has a science of its own – sindonology. This book will look at some of the recent scientific papers published and their conclusions.

It is particularly challenging for scientists that the Shroud may be evidence of an event that could be called supernatural – the resurrection of a dead man. This event would have transcended the boundaries of science and nature. The comment was made by a leading scientist in 1989 that, "If a supernatural explanation is to be proposed, it seems pointless to make any scientific measurement on the Shroud at all."[2] The only possible answer to this is that, even if the event itself was of supernatural origin, it must have left natural traces and evidence that can be measured.

The Shroud challenges non-believers and non-Christians. What if it is indeed evidence of the resurrection of Jesus Christ? What are the implications of this for those who do not believe that Jesus was the Son of God? In April 1997 a fire broke out in the Royal Chapel in Turin Cathedral, where the Shroud was kept. This fire was deliberately started and would appear to have been a deliberate attempt to physically destroy the Shroud. That the Shroud was not burned to cinders was due entirely to the bravery, and faith, of a Turin fireman who went into the burning and smoke-filled Chapel and rescued the Shroud. One can only speculate that one individual in particular felt so challenged by the Shroud that he had to try to destroy it.

Introduction

The Shroud also challenges believers. It is one thing to read the description of Jesus' Passion in the Gospels; it is something else to contemplate the physical evidence of it and to try to understand the effect on a human body of scourging and crucifixion. The Shroud tells the believing Christian that the Passion was no simple execution; it was an experience of suffering, agony and pain almost beyond our comprehension. Anybody who has sat through Mel Gibson's film *The Passion of the Christ* may just start to understand. If one man suffered so much, what was it for? For a Christian that question is answered by his faith.

I approach the topic of the Shroud from the point of view of a qualified scientist, an amateur historian and a believing Catholic. As a scientist I am not intimidated by the scientific papers that have been published on the Shroud. I understand the principles and the science involved. As a scientist I try to keep an open and impartial mind. If there is evidence that the Shroud is not of first-century middle-Eastern origin, let it be studied and judged on its merits. Similarly with evidence that the Shroud is of first-century middle-Eastern origin. I am full of admiration for scientists of other faiths, including those who participated in the STURP team, who have maintained a scientific and scholarly detachment in their studies of the Shroud.

The history of the Shroud, both known and speculated, fascinates me as an amateur historian. Since my school days I have been interested in classical and mediaeval European history. At school I read Gibbons' *Decline and Fall of the Roman Empire* from beginning to end. I still have a quiet wish that the Byzantine Empire had survived until the present day – it would have been fascinating. The history of the Shroud, as will be shown in this book, passes over the highways and byways of both European and middle-Eastern history for the last 2 000 years.

As a Catholic I see the Shroud as the most powerful relic of the birth of Christianity. It has become fashionable to sneer and scoff at relics, and there have indeed been many dubious claims and obvious flights of fancy to gull the naïve and superstitious over the ages. That does not detract from the significance of real relics, of which a number still exist. Others besides the Shroud will be referred to in this book. These relics have both religious significance and historical value. They remind us of historical events and often throw light on these events where there would be no light otherwise. The Shroud in particular falls into this category.

One may ask, why another book on the Shroud? I don't have any remarkable new or fanciful theories as to its origin. However there has been a lot of research and speculation on the Shroud published in recent years. Much of this does not go far beyond conferences held purely to hear these papers; the audience is limited. I believe that there is a need for this latest research and speculation to reach a wider audience. I also feel that the Shroud continues to pose a challenge to everyone who comes into contact with its existence – scientists and non-scientists, the faithful and the sceptical, Christians and non-Christians. This challenge needs to be recognized again and again.

Perhaps the greatest challenge of the Shroud is the fact that it cannot under any circumstances be considered anything but of first magnitude importance. To quote the American writer John Walsh, who is in turn quoted by Ian Wilson in his book *The Blood and the Shroud*, the Shroud is either "one of the most ingenious, most unbelievable products of the human mind and hand on record" or "the most awesome and instructive relic of Jesus Christ in existence".[3] Either way, as a forgery or as a relic, it is a unique challenge to the modern day.

PART 1 –
INTRODUCTION TO THE SHROUD

Chapter 1 – The Shroud of Turin

The Shroud of Turin has been described as the single most studied artifact in history. Whether this is true or not it is certainly one of the most controversial subjects of all time. To the true believer it is the burial shroud of the crucified Christ, left in his tomb at the time of the Resurrection. To the scientist it is an object of study which has yielded a mine of information about subjects as diverse as the medical aspects of crucifixion and the possibility that bacteria can affect the accuracy of radio-carbon dating. The Shroud has given rise to its own branch of science, known as sindonology. To the sceptical it is a piece of mediaeval trickery which has been fooling the gullible for the last six hundred years or more.

The Shroud itself is an ivory-coloured cloth with a herringbone weave. It measures 14 feet 3 inches long by 3 feet 7 inches wide. These measurements may seem a little odd. They make far more sense when converted into first-century Jewish cubits. Using a measure of 21,4 inches to the cubit, based on the Assyrian standard, the measurement of the Shroud converts to exactly 8 cubits in length by 2 cubits in width[1]. It was made in a single piece, apart from a strip approximately three and a half inches wide running the entire length of the left-hand side of the Shroud. This strip is attached to the Shroud by a single seam[2].

On the cloth itself is a faint image, almost shadow-like. This shows the back and front of a well-built man, nearly six foot tall, with a beard and long hair, laid out with his hands crossed in front of him. He appears to be dead, and somehow there is a peacefulness and serenity about his features. The image has been described as "pale and subtle" and "pure sepia monochrome" in colour[3]. There is no visible outline of the image; it melts away into the fabric. It can only be seen clearly from a distance; when viewed from close up it almost seems to disappear.

Also apparent on the Shroud are what seem to be bloodstains. There are flows from several points on the upper forehead as well as from the back of the head; flows from the wrists and the feet; and a copious flow from an elliptical-shaped wound on the left side of the body[4].

The Shroud material is disfigured by stains and by fire damage. One night in December 1532 a fire broke out in the Sainte Chapelle, Chambery, in south-eastern France, where the Shroud was then being kept. The flames rapidly consumed the furnishings in the church and a stained-glass window depicting the Shroud melted in the heat. The Shroud itself was saved by the quick thinking of one of the Duke of Savoy's counselors, Philip Lambert, who, with the help of two Franciscan priests, retrieved the casket containing the Shroud and carried it out of the burning building. However a drop of molten silver fell on to the linen inside the casket, resulting in scorching of all forty-eight folds of the Shroud. This was then doused with water, which resulted in further stains. Almost as if by a miracle the image itself was scarcely touched[5].

The Shroud has been at hazard from fire on at least two other occasions. Framing the thighs of the Shroud's front and reverse body images are four sets of holes and burn marks, three of which appear in right-angle patterns similar to the English letter "L". If the Shroud is folded once lengthwise and once widthwise, all four sets of holes and burn marks superimpose upon one another in the centre of the folded cloth. The charred edges of these holes appear blacker than the fire damage of 1532 and show evidence of the presence of pitch. These holes can be identified in a painted copy of the Shroud dated 1516 and kept in the archives of a church in Lierre, Belgium[6]. Clearly these holes must pre-date the 1532 fire by at least sixteen years, and probably much longer. Their cause is not known, but one suggestion that has been made is that a hot poker was thrust into the Shroud at some time. Hence the holes are known as the "poker holes". They are an important piece of evidence in trying to trace the history of the Shroud prior to its first recorded historical appearance in Lirey, in the Champagne region of France, in about 1355.

Following the 1532 fire the Shroud was sent to a convent of Poor Clare nuns for repair. The nuns backed the Shroud with a simple piece of Holland cloth of the same size, in order to give it strength. They also sewed fourteen large triangular-shaped patches and eight small ones, all made of altar cloth, over the worst of the damage. Other, minor repairs were made in the late seventeenth century, and a few further repairs were made in 1868 by Princess Clotilde of Savoy, who also provided the crimson silk lining cloth on the Shroud[7].

There is evidence of other repairs having been made to the Shroud during the sixteenth century. In 1988 Prof William Meacham of the University of Hong Kong suggested that part of the Shroud had been re-woven in the area from where samples had been taken for radio-carbon dating[8]. It has also been noted that Margaret of Austria, Duchess of Savoy in the early sixteenth century, wanted to leave a portion of the Shroud to her church and stipulated this in her will in 1508. Such a portion may have been removed at the time of her death in 1531 and replaced with a sophisticated patch repair[9]. (This possibility will be discussed in detail in Chapter 18.)

The image on the Shroud shows brutally-inflicted injuries. In addition to the injuries related to the blood flows there are, superimposed on the image in a darker tone of yellow-sepia, more than a hundred dumb-bell shaped markings, each about one and a half inches long. On the reverse image they cover the entire back of the body as well as the cheeks of the buttocks and extend down the legs as far as the ankles. On the frontal image they can be seen on the chest and thighs[10]. They are consistent with the man on the Shroud having received a severe whipping using thongs with dumb-bell shaped metal pellets placed at the end of each thong.

While the image on the Shroud as seen with the human eye gives clear evidence of all of these injuries, it was only with the advent of the photographic age that the true extent of the injuries to the man on the Shroud became clear.

In 1898 a special exposition of the Shroud was held in Turin, to coincide with celebrations of the fiftieth anniversary of the Italian constitution. Although the King of Italy, the legal owner of the Shroud at the time, was reluctant to permit it, a decision was made to allow an amateur photographer with twenty-five years' experience, Secondo Pia, to photograph the Shroud. Despite the primitive state of the photographic technology with which he had to work, Pia succeeded in making two plates. On developing the plates Pia was struck by awe. Instead of the shadowy figures on the

Shroud there was now natural light and dark shading, giving relief and depth. Bloodstains and injuries could be clearly seen. The face, instead of being ethereal, appeared strikingly lifelike against the black background of the negative[11]. What Pia had discovered, quite accidentally, was that the Shroud is itself a negative. A photographic negative of the Shroud has all the properties of a positive photograph.

Pia's negative was of poor quality by modern standards. More modern photographs were taken in 1931 when the Shroud was exhibited to celebrate the marriage of Prince Umberto of Italy. The photographer was a professional, Commander Giuseppe Enrie, and among those attending the photographic sessions was Secondo Pia himself, by then a man in his seventies. Enrie's photographs are of superb quality. Further photographic sessions were held in 1969 and 1973[12], under rigidly controlled scientific conditions

The photographic negatives clearly show four categories of injuries on the man on the Shroud. Firstly there are the injuries as from a severe whipping. Secondly there appear to be injuries from various forms of incidental abuse, such as bruising to the right eye and both cheekbones and the nose possibly being broken, consistent with a severe assault. The shoulders appear chafed as if from carrying some heavy object and the knees seem to have been damaged by falls. There are signs of sharp objects having been forced into the man's head, resulting in numerous bloodflows. Thirdly there are injuries to the wrists and feet as if they had been pierced in some way. Finally there is the injury to the chest which would appear to have been caused by a bladed weapon[13].

It does not take a great leap of imagination or faith to realize that these injuries coincide with those described in the Gospel accounts of the Passion and death of Jesus Christ.

* * * * * * * * *

Since Secondo Pia's first photographs the Shroud has been the subject of serious scientific study. Early studies were carried out at the start of the twentieth century by Paul Vignon, a professor of biology at the Institut Catholique in Paris. His first conclusion was that the image on the Shroud was not a painting. However a presentation of his findings at the French

Academy of Sciences in 1902 by his colleague Prof Yves Delage met with harsh criticism[14]. French scientists at the beginning of the last century were not prepared to face the challenge posed by the first photographs and early research on the Shroud.

Modern study of the Shroud started in the 1970s. In 1973 threads were taken from the Shroud for study and studies were also made of the photographs. Studies were made of the image itself. However the report from these studies was inconclusive[15]. On the other hand they did open the way for the major research carried out a few years later by the Shroud of Turin Research Project (STURP). This brought together a group of scientists prominent in their various fields who were able to carry out a detailed study of the Shroud over 120 continuous hours in 1978[16]. They concluded that no pigments, paints, dyes or stains could be found on the threads of the Shroud. They further concluded that:

"There are no chemical or physical methods known which can account for the totality of the image, nor can any combination of physical, chemical, biological or medical circumstances explain the image adequately.... We can conclude for now that the Shroud image is that of a real human form of a scourged, crucified man[17]."

The one significant scientific test that the STURP team did not carry out was a radiocarbon dating test. In 1988 three samples from the Shroud were subjected to radiocarbon testing at three laboratories in Arizona, Oxford and Zurich. Their conclusion is well-known – they dated the material of the Shroud to between 1260 and 1390, coinciding with its first recorded appearance in history at Lirey in France. This resulted in an immediate response in the international press that the Shroud had been proved to be a mediaeval fake. The result of the radiocarbon testing was accepted with serenity by the Archbishop of Turin, Cardinal Anastasio Ballestrero[18]. However many believers in the authenticity of the Shroud were shocked by this result and some refused to accept it. Allegations were even made that the scientists involved in the testing had been dishonest – something that no serious researcher on the subject has ever suggested.

The radiocarbon dating of the Shroud flew in the face of the results of the STURP researchers. Over the following fifteen years theories were developed to try to reconcile the radiocarbon dating results with a true age of the Shroud of 2 000 years. An American microbiologist, Dr Leoncio Garza-Valdes, carried out research suggesting that the fibres of the Shroud may have been contaminated with other organic material produced by bacteria and that this contamination is responsible for the incorrect result of the radiocarbon dating[19].

It has also been suggested that the samples used for the radiocarbon dating were contaminated by material from a patch in the Shroud that had been inserted during repairs in the sixteenth century, but which had been inserted so subtly that it could not be distinguished from the original Shroud material surrounding it[20].

Support for this suggestion has been given by Dr Ray Rogers of the University of California, a former member of the STURP team. In 2005 he published a scientific paper in which he showed that he had dated the Shroud to between 1 300 and 3 000 years old by a chemical analysis method[21]. Dr Rogers was quoted by the BBC as saying that "the radiocarbon sample has completely different chemical properties than the main part of the Shroud relic" and that his research and chemical tests show that the sample used in the 1988 radiocarbon analysis was cut from a mediaeval patch woven into the Shroud, in his opinion to repair fire damage[22].

Samples of blood have also been taken from the Shroud. These have been analysed and determined to be the blood of a human male[23]. Further, the blood has been determined to be of the AB group. In the general world population this blood group is relatively rare (3,2 per cent of the population). However its highest incidence (18 percent of the population) occurs among the "Babylonian" Jews and the Jews of northern Palestine[24]. The samples were not sufficient for testing of the rhesus factor.

Later chapters will cover the scientific studies of the Shroud in much greater detail.

* * * * * * * * *

A major concern in recent years has been the conservation of the Shroud, to ensure that it does not deteriorate. This was referred to by Cardinal Ballestrero in an interview he gave in 1988, in the course of which he said:

"The present problem is to provide an adequate method for the conservation of the Shroud, a method that will guarantee the cloth the greatest possible security[25]."

In 2002 a conservation programme was carried out by a Swiss textile conservator, Mechthild Flury-Lemburg. The backing cloth and the patches were removed and the Shroud was sewn on to a fresh backing cloth. This programme led to much criticism and allegations that the advice of Shroud experts had been ignored and that potential vital microscopic evidence had been destroyed. The Shroud author and historian Ian Wilson was given the opportunity to review the restoration work that had been done. In a commentary entitled "The New, Restored Turin Shroud" he expressed himself relieved to find the alarmist allegations totally unfounded and extended his warmest congratulations to the restorers for their endeavours. He also expressed his confidence that the Shroud has never been in more capable or caring hands than it is now[26]. There are other researchers, however, notably Prof William Meacham, whose views are diametrically opposed to such expressions of confidence.

Despite the detailed scientific studies of the Shroud over the last forty years, theories, some serious and some fanciful, continue to abound as to how the Shroud originated in mediaeval times. This is part of the Shroud's challenge – if its authenticity as the burial cloth of Christ is too hard to accept, what else could it be? It has been suggested that it is itself a photograph, produced by primitive photographic methods, possibly even by Leonardo da Vinci[27]. It has also been suggested that the image on the Shroud is that of Jacques de Molay, the last Grand Master of the Templars, resulting from torture while he was imprisoned by King Philip IV of France[28]. Another scenario that has been suggested is that the Turks crucified some unfortunate Crusader towards the end of the Crusades, as a mockery of the Passion of Christ, and the Shroud is actually the burial cloth of this unfortunate[29]. There are others who still maintain that it is a painting despite all the evidence to the contrary. The letter of Bishop d'Arcis of Troyes addressed to Pope Clement VII in 1389, in which he described the Shroud as having been "cunningly painted"[30] has a lot to answer for!

Undoubtedly the Shroud will continue to generate controversy and to challenge those who enquire about its nature and origins. Many questions about the Shroud have been answered in the last forty years. However, many more remain, not least the question of how the image was actually formed. Many of these questions, including this question of image-formation, will be addressed in subsequent chapters.

PART 2 –
THE SHROUD AND HISTORY

Chapter 2 – Jerusalem, Antioch and Edessa

In August 944, in the imperial capital Constantinople, a sermon was preached by the Archdeacon of the Church of Hagia Sophia, Gregory Referandarius. This was no ordinary occasion. Gregory's sermon was delivered to mark the arrival in Constantinople of the Image of Edessa, an image on a cloth described by a Greek word *acheiropoietos* meaning "not made by human hands". This sermon has recently been translated in full into English[1].

In his sermon Gregory told the story of Abgar, king of Edessa during the lifetime of Jesus Christ. According to this story Abgar, who was suffering from some disease, wrote to Jesus to say that he had heard about the cures Jesus was performing in the region of Jerusalem and to ask him to come to Edessa to cure his illness. Jesus replied that when he had "been taken up" he would send one of his disciples to "cure your sickness and bring life to you and those with you".

Gregory went on to say that he had traveled personally to Edessa to discover more of this story. In Edessa, he said, he had found a number of manuscripts in Syriac, which went on to tell how the disciple Thaddaeus had traveled to Edessa after the Resurrection and Ascension of Jesus. Thaddaeus had brought with him a linen cloth with an image on it and this cloth had cured Abgar's disease. In mentioning the linen, Gregory used the Greek word *othone*, which is the singular of the word *othonia* used in Luke's Gospel to describe Christ's burial cloths[2]. Thaddaeus told Abgar that Jesus had miraculously imprinted the reflection of his form on the linen.

The story of Thaddaeus's mission to King Abgar and Edessa is an old one. It is recorded in detail in an ancient Syriac document, *The Teaching of Addaeus the Apostle*[3]. The story is also repeated by Eusebius Pamphili, Bishop of Caesarea, in his *Ecclesiastical History*[4]. Eusebius refers to the correspondence between Abgar and Jesus as being in the archives at Edessa, and the translator of his *History* has noted that Eusebius in all probability actually found these epistles – although he (the translator) describes them as apocryphal – in the public archives there[5]. A further witness to their existence is a fifth century Armenian historian, Moses Chorenensis. Whether or not these letters were genuine or merely later forgeries is not really important – Eusebius clearly believed that they were genuine. These letters also found their way into English religious practices. In the British Museum there is a very old service-book dating to Saxon times in which the letter from Jesus to Abgar follows the Lord's Prayer and the Apostles' Creed. The story of Abgar is also told in an Anglo-Saxon poem. Reverence of the letter continued in parts of England at least until the nineteenth century and may still continue today in a few places. The letter was recognized as the word of God and a genuine epistle of Christ[6].

According to the story, as recounted both by Eusebius and in the Syriac document, Thaddaeus healed Abgar of his disease "without drugs and herbs". He also healed many others. His arrival in Edessa is dated as being in the "340th year", which equates to 30 AD. Thaddaeus had been sent to Edessa by the Apostle Thomas.

The first part of *The Teaching of Addaeus the Apostle* is taken from a manuscript of the fifth century and the rest from a manuscript of the sixth century. However it is clearly a document from an earlier date. Another early Syriac document, *The Teaching of Simon Cephas,* is thought to have been written in either the late first or early second century[7]. Yet another document, *The Teaching of the Apostles,* is equally old. Interestingly, this latter document refers to there being "Guides and Rulers" in the churches after the deaths of the Apostles[8]. This term is believed to have pre-dated the use of the term "Bishop" for an ecclesiastical ruler and is also found in *The Teaching of Addaeus the Apostle*, thus giving an indication of a very early date for this document as well.

There is also a reference in *The Teaching of the Apostles* to Thaddaeus going to Edessa. In listing which regions were evangelized by which apostle or disciple, this document says:

"Edessa… received the apostles' ordination to the priesthood from Addaeus the apostle… who was priest and ministered there in his office of Guide which he held there[9]."

Finally, as a further indication of the early evangelisation of Edessa, another Syriac document, *The Acts of Sharbil,* suggests that in the reign of the Emperor Trajan (AD 98 – 117) paganism and Christianity were both tolerated in Edessa[10].

The story of Thaddaeus's mission to Edessa immediately after the death of Jesus is therefore historically credible. However, none of these very early documents makes any mention of a miraculous image, cloth or shroud. Neither does Eusebius. Eusebius would certainly not have mentioned anything about an image. The early Christians of the first and second centuries shared the Jewish abhorrence of imagery. It was only after the conversion of the Empire under Constantine that images of Jesus and Mary started to circulate and become fully acceptable. Even then a number of Church leaders continued their strong disapproval. Eusebius was one of these. He is said to have rebuked Constantia, the half-sister of the Emperor Constantine, for her idolatry when she wrote to him asking him to obtain for her a picture of Christ[11].

There is a late fourth century manuscript preserved in St Petersburg, entitled *The Doctrine of Addai*[12]. This appears to be a corrupted version of *The Teaching of Addaeus the Apostle*, with a large number of anachronisms and a particular reference to Hanan, a messenger from Abgar to Jesus, painting Jesus's likeness "with choice paints". The fact is often overlooked that Jesus, his family and his immediate followers were all Jews who followed Jewish customs and Jewish religious practices. The idea of Jesus the Jewish teacher permitting a pagan to paint his likeness is utterly unthinkable. The association of Thaddaeus with a cloth with an image on it – whether miraculous or painted – is clearly a later story dating from the time when religious imagery was becoming acceptable.

We therefore have two separate stories and traditions – a very early one, and probably authentic, of Thaddaeus the disciple going to Edessa on a mission of evangelisation, and a later, more dubious one that involved Thaddaeus with a cloth with an image on it.

*   *   *   *   *   *   *   *   *

If it is accepted as possible that a shroud or other cloth associated with the Passion and death of Jesus Christ survived and eventually became the Cloth of Edessa, as suggested by Gregory Referendarius, then it is necessary to look at the whole issue of relics and attitudes towards them. The word relic is derived from the Latin *reliquiae,* meaning remains, and in religious terms it refers particularly to the physical remains of a holy person or items closely associated with that person. The veneration of relics is an ancient practice, associated with both Christians and non-Christians[13]. Relics of Buddha, who died in 483 BC, were distributed soon after his death and a limited number of authentic parts of his body, including teeth and hairs, remain in existence today. There are two hairs of Mohammed, who died in AD 632, preserved in a reliquary at the Dome of the Rock in Jerusalam. The Old Testament makes reference to relics of the prophet Elisha:

"Elisha died, and was buried. Now bands of Moabites were making incursions into the country every year. Some people happened to be carrying a man out for burial; at the sight of one of these bands, they flung the man into the tomb of Elisha and made off. The man had no sooner touched the bones of Elisha than he came to life and stood up on his feet[14]."

Relics of the life of Jesus and particularly his Passion and death would have been revered by the early Church. It must be considered unlikely in the extreme that any major relic of Jesus would have been given to such an obscure disciple as Thaddaeus and immediately taken to a foreign land such as Edessa. An alternative hypothesis for the early history of what became the Cloth of Edessa has been suggested[15].

Firstly, it is necessary to imagine the events of that first Easter morning. St John's Gospel describes how Mary Magdalene came to the tomb early in the morning and found it empty. She immediately ran to tell Peter and John. Her first thought was the obvious one, someone's taken the body. This must have been a considerable shock and the Gospel tells how she wept because of it. Peter went into the tomb where he saw the linen cloths on the ground and, apart from them, the cloth that had been over Jesus's head[16]. Although the Gospel does not say as much, what would have been more natural than for Peter and John to take the cloths away with them, to ensure that they were not stolen and for the purpose of remembrance?

As the Church developed in Jerusalem in its very early days these cloths would have been valued. To describe them as objects of reverence would be to ignore the Jewish prohibition against graven images and against worshipping anything man-made. The burial cloth or shroud could have caused serious problems at this time if it had indeed borne the image of the crucified Christ. It would have had great value as a memento of the Crucifixion and Resurrection but would at the same time have been anathema as an image. It would undoubtedly have been a source of great soul-searching among the Apostles. It could not be destroyed because of its value; the best thing might well have been to hide it away, out of sight and out of mind. It is possible that at some stage – probably later rather than earlier – a distinction came to be made between a graven, or man-made, image and the miraculous image on the linen cloth, not made by human hands – hence the Greek term *acheiropoietos* subsequently applied to the Cloth of Edessa.

The Acts of the Apostles tells how the Church in Jerusalem was persecuted following the martyrdom of Stephen[17]. Some of the Christians who left Jerusalem at that time, particularly those originating from Cyprus and Cyrene, traveled to Antioch and started preaching to the pagans there. As a result Barnabas was sent by the Church in Jerusalem to Antioch[18]. One hypothesis that has been suggested is that relics of Christ's Passion were taken from Jerusalem to Antioch at this time, possibly by Barnabas but more likely by Peter[19]. This would have been a very convenient way of getting a relic that was slightly embarrassing away from the centre of Judaism and into the hands of pagan converts who would not have been quite so sensitive about images. If it was taken to Antioch it could have been concealed until at least the middle of the fourth century. By this time images and depictions of Jesus and Mary had become more accepted by the Christian Church and there would have been no further embarrassment about a valuable cloth with an image on it. During these three centuries Antioch had also been the centre of numerous persecutions of Christians by the Roman authorities. This would have been another reason to keep a valuable relic hidden, to prevent its seizure and destruction by the persecutors.

However, there is yet another possibility which may be the most likely of all. What would have been more natural than to give the relics of the

Passion into the keeping of their rightful owners, the family of Jesus? It is often overlooked that the early Jerusalem church was led by Jesus' immediate family. Following the departure of Peter from Jerusalem in about 42 AD, leadership of the Jerusalem church passed into the hands of James, known as "the brother of the Lord". In addition, Jesus' mother Mary almost certainly remained in Jerusalem for the remainder of her life and must have been a venerated figure among the Christian community there. Following the martyrdom of James in 62 AD leadership of the Church in Jerusalem passed to Symeon, a cousin of Jesus by virtue of his being the son of Clopas, the brother of Joseph[20]. Symeon led the Church into temporary exile at the time of the Jewish revolt of 66 – 70 AD. The Jewish Christians refused to take part in this revolt, with the result that they, unlike other Jews, were allowed to return to Jerusalem after the revolt had been put down. They returned to their former quarter on Mount Zion and resumed their way of life[21].

Despite this initial clemency to Jewish Christians, during the next sixty years the Jews of Palestine were treated severely by the Roman authorities. They were particularly persecuted during the reigns of the Emperors Trajan and Hadrian, who regarded them as an extremely troublesome people. The outcome of this was the final rebellion of the Jews under the leadership of a man called Bar Chochebas. The Christians in Jerusalem refused to join this rebellion, as they had refused to join the earlier Jewish War, with the result that they were cruelly treated by the rebels. The rebellion was finally crushed by the Roman legions of the Emperor Hadrian in 135 AD. Jerusalem was totally destroyed and renamed Aelia Capitolina. The Jews were dispersed and no Jews were permitted to remain in Jerusalem. This time this was applied to the Jewish Christians as well as to other Jews[22].

The Jewish Christians in exile developed into different groups. The most important group was known as the Ebionites, possibly from the Hebrew word *'ebjon*, meaning poor. The Ebionites seem to have separated into two distinct groups about the year 150 AD – one group saw Jesus as a mere man and the other continued to acknowledge him as the Messiah and Son of God. Some Ebionites eventually adopted Gnostic views. Their later history is uncertain[23].

Eusebius lists fifteen Jewish Christian bishops in Jerusalem prior to the destruction in 135 AD – of these only two are known to history – James

Jerusalem, Antioch and Edessa

and Symeon[24]. The date of Symeon's death is unclear. One tradition has it that he was crucified during the reign of Trajan at the age of 120. This advanced age seems most unlikely, and there would not have been a lot of time for another thirteen bishops between his death at such a late date and the destruction of Jerusalem. His death can probably be put at a considerably earlier date.

Eusebius also relates another story that stresses the importance of the family of Jesus in the early church. The Emperor Domitian at the end of the first century instituted a persecution of both Christians and Jews. Two Jewish Christians were brought before the Emperor to be questioned about their beliefs – they were the grandsons of Jude, the brother of James and hence also a "brother of the Lord". These two men were small farmers. They explained to the Emperor that their kingdom was a heavenly one. Domitian accepted their explanation and released them. He also ended his persecution. According to Eusebius the two men lived on into the reign of Trajan and guided the churches. The story is almost certainly apocryphal – one cannot imagine a Roman Emperor summoning two peasants to explain their beliefs to him and then ceasing a persecution solely on the strength of their explanation. However the story does emphasise once again the importance that the early Church gave to Jesus's immediate family[25].

There is no reason at all why the Shroud should have left Jerusalem while Jesus's close relatives such as James and Symeon remained prominent in the Jerusalem church. After the death of Symeon, at whatever date that was, the situation may have changed. In view of on-going persecutions – and the Romans of the time are unlikely to have made much distinction, if any, between Jewish Christians and orthodox Jews – a decision may have been made to move valuable relics that were easily portable to some other place of safety. Antioch would have been a likely place for such a purpose. It was the third city of the Roman Empire, after Rome itself and Alexandria. It had been a centre of Christianity since the earliest days of the Church. It was also not all that far from Jerusalem.

From this time on the Shroud may well have been kept in hiding in Antioch, to protect it against both Roman persecutors of the Christian Church and iconoclastic Jews and Christians[26]. The aversion of many Christians, particularly in the Eastern Church, to images continued even after the conversion of the Empire to Christianity under the Emperor Constantine the Great. The heresy of Arius, who taught that Jesus was not truly God, in the fourth century had a pronounced effect on Church affairs in Antioch. The Church there was divided between Arian and orthodox factions, each with its own bishop, for most of the fourth century. The Shroud, possibly with other Passion relics, may have come into the possession of the Arian faction in Antioch[27].

In 357 AD the Arian faction achieved predominance in Antioch and obtained possession of the Golden Basilica of Constantine, the cathedral in that city. It may have been then that the Shroud was exhibited for the first time, within the confines of the cathedral and to the Arian faithful[28].

It is possible to relate this date to the first appearances of Shroudlike representations of Jesus. The very earliest depictions of Christ show him as youthful and beardless. Such depictions have been found at a very early date in Asia and slightly later in Rome, England and Ravenna[29]. One such example is a mural of Christ and the Apostles in Milan dating from the period 355 – 397 AD[30]. The youthful, beardless image of Jesus also appears on early sculptures in the south of France and Gaul and continues up to the early years of the Holy Roman Empire in the ninth century[31]. Today there is a mural of Jesus as the Good Shepherd in Lincoln Cathedral, England, showing him clean-shaven and youthful. This clearly seems to have been inspired by the early depictions in the Roman catacombs and elsewhere.

On the other hand there were some bearded representations of Jesus in the West. One example is a mural of Christ and the Apostles at Sta Pudenziana in Italy dating from 402 - 417. However in this mural, although appearing bearded, the figure of Christ bears no resemblance to the Shroud figure[32]. What seems likely is that representations of Christ in different regions reflected the contemporary appearance of people in those regions. Thus in the Romanised West a clean-shaven appearance would have been most common. In the East and the Hellenistic world, where it was the custom to wear a beard, bearded representations of Jesus would have been standard. There were of course always exceptions.

However in the late fourth century depictions of the classic Shroudlike Christ began to be depicted in sarcophagi now to be found in Italy[33]. Various sarcophagi of the Theodosian era (370 – 410) portray Christ with a narrow and majestic face, a medium-length beard and long hair that falls on the shoulders, sometimes with a centre parting. These sarcophagi include a fragment in St Sebastian Outside the Walls in Rome, dated to around 370, a sarcophagus in Arles pre-dating 370, one in the church of St Ambrose in Milan dated 380 – 390 and the Sarcophagus #151 of the Lateran collection in the Vatican, dated close to 400. In all of these works the face of Christ corresponds in all its essential traits to the face on the Shroud[34].

Further, whereas the depiction of Christ as a youthful, beardless figure continued in the West for some centuries, whenever Christ was depicted in the context of the Passion it was with the Shroud-like features that first appeared in the fourth century. In the East this type of Christ became the norm in all figurative art[35].

This may be evidence that the Shroud had been placed on display, wherever it was, and that the image had had an influence on artists depicting the face of Christ. This would have been an indirect influence, as Western artists in particular would not have seen the Shroud, which was in the East, but may have seen representations of it. It has to be said, however, that it was only in the middle of the sixth century, by which time the Cloth of Edessa was known, that images of Jesus appeared in which specific correlations can be made between points on the images of the time and points on the image on the Shroud. These have been referred to as "spy" elements, which confirm the dependence of any particular work of art on the Shroud[36].

It was also about the time of these very early depictions of the Shroud-like Christ that the legend appeared of King Abgar of Edessa having obtained a painting of Jesus's likeness, made by his court painter Hanan, following which he was visited by the disciple Addai who cured him of an incurable disease. This legend comes from the corrupted document *The Doctrine of Addai*, which, as mentioned earlier, dates to around the beginning of the fourth century[37]. This is the earliest connection between Edessa and a possible image of Jesus and may also be an indication of the public appearance somewhere of what later became the Cloth of Edessa.

At the time of the Emperor Julian (361 – 363), known as the Apostate because of his rejection of Christian beliefs, certain valuable objects held at the Great Cathedral of Antioch appear to have been hidden and they may have been lost for some period of time. The Treasurer of the Cathedral, Theodoretus, suffered torture and death rather than reveal the whereabouts of these objects to the Emperor's agent[38]. If the Shroud was indeed in Antioch it is likely that it would have been one of the objects hidden at that time.

The Arians were expelled from Antioch following the establishment of Catholic orthodoxy by the Emperor Theodosius in 380 AD. Doctrinal differences, however, continued to affect the Asian church until the early sixth century, with the Monophysites, who believed that Christ only had a single, divine nature, predominating in Antioch. In 526 and again in 528 Antioch suffered severe damage from major earthquakes. The following year the Kingdom of Persia threatened to attack Antioch. Subsequently a truce was reached between Persia and the Empire, but Persia finally attacked and destroyed Antioch in 540[39].

The history of the Shroud contains numerous instances where the Shroud was moved to keep it safe from invasion, unrest and disorder in general. This would seem likely to have been one of those times. In any event, a holy icon, to become known as the Cloth of Edessa and described as "not made by human hands", was reported by the Syrian historian Evagrius, writing late in the sixth century, to have been present in Edessa during a siege by King Chosroes of Persia in 544[40]. This is the earliest documentary evidence of the existence of the Cloth of Edessa – all hypotheses or attempts to explain its history prior to this, however attractive they might sound, remain in the sphere of speculation.

Edessa is about 145 miles north-east of Antioch and was also a centre of the Monophysites. With Antioch devastated by natural disasters and threatened by military attack, it can be understood that any valuable relic, such as the Shroud, would have been moved to a place of greater safety. Edessa would clearly have been a logical choice for such a place of safety, being both fortified and in the hands of co-religionists.

During Chosroes' siege of Edessa he constructed a timber siege tower from which missiles could be fired down into the city. The Edessans sought to dig a mine underneath the tower and set fire to it. This plan initially failed. According to the legend of the Cloth of Edessa, the inhabitants of the besieged city then brought the Cloth forward. The wood in the mine caught fire, the Persian siegeworks collapsed and the Persians shortly thereafter abandoned the siege. In this way the Cloth gained recognition as both a holy relic and an effective palladium against hostile attacks.

Evagrius described the events of the siege and the intervention of the Cloth in the following terms:

"The mine was completed; but they (the Edessans) failed in attempting to fire the wood, because the fire, having no exit whence it could obtain a supply of air, was unable to take hold of it. In this state of utter perplexity they brought out the divinely made image *not made by the hands of man* (author's italics), which Christ our God sent to King Abgar when he desired to see him. Accordingly, having introduced this sacred likeness into the mine and washed it over with water, they sprinkled some upon the timber… the timber immediately caught the flame, and being in an instant reduced to cinders, communicated with that above, and the fire spread in all directions.[41]"

This would also seem to be the earliest reference connecting a King Abgar of Edessa directly with Christ and an image not made with hands, as opposed to the painting made by the court painter Hanan, which is referred to above. This was the legend that developed and which was referred to in detail some three centuries later by the Archdeacon Gregory Referandarius of Constantinople in the sermon he gave on the occasion of the arrival of the Cloth of Edessa in Constantinople, referred to at the opening of this Chapter.[42]

* * * * * * * * *

It has been suggested that, at this time, the Cloth suffered some fire damage. In order to conceal this damage the Edessan church hierarchy folded the cloth in four to create the portrait known as the Image of Edessa[43]. This four-folded nature of the Shroud has been the subject of much speculation. There are two very early references to either the Cloth of Edessa or a similar cloth having been folded in four, specifically using the unusual Greek word *tetradiplon*. A late sixth-century document called the *Acts of Thaddaeus* makes the following reference to an image of Christ on a cloth folded in four:

"And he (Christ)… asked to wash himself, and a "doubled-in-four" (*tetradiplon*) cloth was given to him: and when he had washed himself he wiped his face with it. And his image having been imprinted upon the linen…[44]"

One year after the arrival of the Cloth of Edessa in Constantinople, in 945, a special Official History of it was written[45]. This includes in detail the legend of King Abgar and his cure by a miraculous image of Christ. The original Greek text of this History uses the word *tetradiplon* once in its description of the Cloth[46].

It can be speculated therefore that the Cloth of Edessa, or the Shroud of Turin if the two are indeed the same article, was folded in four at an early stage in its history. However it should also be noted that the "poker holes" in the Shroud of Turin, which have been referred to in Chapter 1, were clearly made when the Shroud was folded once lengthwise and once widthwise. (In connecting the Shroud with the Cloth it is also worth noting that the "poker holes" can be seen on a shroud in a twelfth century manuscript, the Hungarian Pray Manuscript, which depicts the dead Christ prior to burial. The Pray Manuscript and its importance in Shroud history will be covered in more detail in the next Chapter.)

If the Shroud is identical with the Cloth these holes must have been made before it was folded in four, or no later than the sixth century. If this is the case it may not be unreasonable to connect the fire damage that caused the "poker holes" with the Cloth's exposure to fire at the siege of Edessa in 544.

It would be miraculous indeed if wood that had failed to burn because of lack of oxygen burst into flame upon being sprinkled with water. It has been suggested that this was a story concocted at the time to conceal a more mundane truth – that the sacred cloth had suffered fire damage through the carelessness if not the deliberate action of its Edessan custodians. This fire damage may have included the "poker holes", possibly the result of a pitch-soaked firebrand being applied to the exact centre of the Cloth in some sort of ritual aimed at igniting the wood in the mine.

Having seen the extent of the damage to the relic, the Edessans then sought to hide it by folding the Cloth in four[47].

This is of course purely speculative. The origin of the "poker holes" in particular can never be known with any certainty. As speculation, however, it does not contradict anything that is known about the Shroud of Turin, and it is indeed part of a pattern that links the modern-day Shroud of Turin with the Byzantine Cloth of Edessa.

There is a further possible explanation for the "poker holes". It is quite understandable for English-speaking writers to describe the pattern of the holes as resembling the letter "L". However, an inverted "L", when reversed, forms the upper-case form of the Greek letter *gamma*.

In the Byzantine era the *gamma* and notched bands of Jewish *talitoth* were used as decorations on tunics and altar cloths. Although the *gamma* was used specifically on the tunics of women, Christians of the time seem to have overlooked this. Around the sixth century, at the time that the Cloth of Edessa made its appearance, the *gamma* marking was used on altar cloths which were known as *gammadia*. It may be that whoever burned the "poker holes" into the Shroud was marking it, in his way, as a holy cloth[48].

* * * * * * * * *

The value placed on the Cloth by the Edessans is reflected in the fact that copies of it began to circulate in the late sixth century. Edessa's neighbouring towns, Melitene and Hierapolis, both acquired copies. In Cappadocia an image of Christ was "discovered" in a water cistern in the town of Camuliana by a woman who, apparently, desired to see Christ. This became a rival palladium to the Cloth of Edessa and was carried around Asia Minor by fund-raising priests before being sold to Constantinople in 574.

At about the same time a description appeared of a linen cloth with an imprint of Christ, quite possibly a copy of the Cloth of Edessa, that had been found at Memphis in Upper Egypt.

Edward Gibbon described this situation in his *Decline and Fall of the Roman Empire*. He wrote that "images made without hands" were circulating in the camps and cities of the Eastern Empire. They were worshipped and were also the instruments of miracles. They also had the power to strengthen the hope and courage of the imperial legions, or, on the other hand, repress their fury[49].

From the middle of the sixth century onwards the image of the Cloth of Edessa became a clear model for the artistic figuration of the image of Christ[50]. Numerous representations of Christ exist from that period, showing features that clearly relate to the features of the image on the Cloth of Edessa. One notable example is the gold *solidus* issued in about 692 by the Emperor Justinian II bearing a portrait of Christ Enthroned. This was the first proper likeness of Jesus to appear on a coin and has been shown to bear striking similarities to the face on the Turin Shroud[51].

In 639 Edessa was captured by Moslem forces. The new Moslem rulers of the city accorded tolerance to the various Christian denominations there and the Cloth of Edessa continued to be venerated as the city's palladium. Despite outbreaks of iconoclasm – the destruction of icons and images – in both the Byzantine and Moslem empires during the eighth and ninth centuries, the Cloth of Edessa remained unscathed.

Further references to the Cloth are known from the eighth century. St John of Damascus writing in his work *On Images* in about 730 referred again to the Abgar legend:

"Furthermore there is a story told about how, when Abgar was lord of the city of Edessenes, he sent an artist to make a portrait of the Lord, and how, when the artist was unable to do this because of the radiance of His face, the Lord Himself pressed a bit of cloth to His own sacred and life-giving face and left His own image on the cloth and so sent this to Abgar who had so earnestly desired it[52]."

St John was writing in defence of the use of images in worship, against the prevailing iconoclastic policy of the Byzantine Empire. The reference to the Abgar legend clearly relates to the Cloth of Edessa. In the context of St John's writing, this reference also shows that the image on the Cloth of Edessa was considered significant in the iconoclastic controversy (which is covered in more detail in the next Chapter.)

St John described the cloth as a *himation*. This is a Greek term referring to an outer garment, similar to a Roman toga[53]. Such a garment would usually be approximately two yards wide by three yards long[54]. This clearly indicates that the full size of the Cloth was known at that time, even if it continued to be folded in four.

A further reference appears to both the size of the garment and the image being miraculously imprinted on the Cloth in the *Good Friday Sermon* made by Pope Stephen III in 769, where he said:

"He stretched His whole body on a cloth, white as snow, on which the glorious image of the Lord's face and the length of His whole body was so divinely transformed that it was sufficient for those who could not see the Lord bodily in the flesh to see the transfiguration made on the cloth[55]."

In 787 Leo, Lector of Constantinople, told the Second Council of Nicaea that he had visited Edessa and seen there "the holy image made without hands and adored by the faithful[56]." This is yet a further indication of the significance of the image on the Cloth of Edessa to those opposed to the iconoclastic views of the Byzantine Emperors of the eighth century. This Council brought an end to the first period of iconoclasm in the Empire.

The Cloth could not remain undisturbed for ever. The time came when it would be moved to Constantinople.

Chapter 3 – From Edessa to Constantinople

Constantinople was founded in 324 by the Roman Emperor Constantine the Great, to be the site of his capital, a New Rome. It was situated on the Bosphorus at the entrance to the Black Sea, on the site of the old Greek town of Byzantium. Today it is the Turkish city of Istanbul.

Constantine intended his New Rome to combine Roman imperial tradition with Christian orthodoxy. Greek culture came to be a third element of the mix. Constantine established a Senate on the model of the Roman Senate, public buildings were erected and imperial bureaucracy duplicated. Aristocratic families from Italy were encouraged to build residences there and even the customs of chariot racing in the Roman circus were transplanted to Constantinople.

At the time of Constantine's reign the Empire had been effectively divided into two administrative portions – East and West. The establishment of Constantinople as the centre of the Eastern Empire formalized this division. Although some of Constantine's immediate successors ruled as sole Emperors of both Eastern and Western Empires, by the end of the fourth century the division was made permanent. The Western Empire ceased to exist in the late fifth century, but the Eastern Empire continued to flourish.

Constantine had founded his city with the intention that it would be a centre of Latinity in the East and indeed in the sixth century Latin was still the native tongue of the Eastern Emperor Justinian the Great. However Greek culture had been predominant in the East for centuries and the linguistic balance in the Empire inexorably tilted in favour of Greek. By the end of the sixth century, according to Pope Gregory the Great, it was no easy matter to find a competent translator from Latin into Greek[1]. Justinian was the last purely Roman-minded Emperor.

The border between the Roman and Persian Empires had been an area of conflict for centuries. In 165 Mesopotamia had been conquered and attached by the Roman Empire. This region included the city of Edessa, which remained part of the Eastern Empire for the next five hundred years. In the seventh century the Emperor Heraclius decisively defeated the Persian Empire, but this triumph was rapidly followed by Muslim Arab expansion and the Arab conquests of Egypt, Palestine and Syria. The Arab armies reached the gates of Constantinople before being turned back.

At the same time the Eastern, or Byzantine, Empire lost its remaining western provinces, including Spain and Africa. From this time onwards the Empire faced east with its heart in Asia Minor.

* * * * * * * * *

Although the Muslim rulers in Edessa may have extended some tolerance to the exhibition of the Cloth in their midst, this tolerance was not always shared by their superior in Damascus, the Caliph. To true Moslems any kind of picture, statue or representation of the human form is an abomination. It can be assumed that this intolerance was extended even to representations of apparently miraculous origin. In the late seventh and early eighth centuries the Caliphs tried to prevent the use of pictures among their Christian as well as Moslem subjects. It must be assumed that, during periods of particular intolerance, the Cloth of Edessa was kept hidden away where it would not cause offence. In any event it survived undamaged.

At the same time there were a number of bishops in the Byzantine Empire who had similar views. The Byzantine Emperor Leo III the Isaurian was influenced both by the Caliph Omar II, who tried to convert him to Islam but who merely succeeded in turning him against images, and by the Christian enemies of images. Leo reached the conclusion that images were the chief hindrance to the conversion of Jews and Muslims, the cause of superstition, weakness and division in his Empire and a breach of the First Commandment. In 726 he published an edict declaring images to be idols and commanding all such images in churches to be destroyed. This was the start of the period of iconoclasm in the Empire, which was marked by serious disturbances and the persecution of monasteries in particular, as monks tended to remain loyal to the old customs of the Church which included the veneration of images. It also led to a major breach between the Emperor and the Pope, as a result of which the papacy gained its independence from the Empire.

The principle of iconoclasm extended further than just to man-made images. Under Leo III relics of saints were destroyed, shrines broken open and the bodies of saints that had been buried in churches were burned. The destruction of relics continued for several decades. It is not unlikely that had the Cloth of Edessa, with its visible image, been in Constantinople at this time it, too, would have been destroyed.

At this time one of the most significant defenders of images was St John of Damascus, who, as we saw in the last Chapter, wrote about the Cloth of Edessa in about 730. Clearly at this time it was on display in Edessa despite the iconoclastic views of the Caliph and the upheavals in the Byzantine Empire.

The persecution of Leo III was continued and even intensified by his successors. It was only after the death of his grandson, Leo IV, that the first period of iconoclasm was finally brought to an end in 787 by the Second Council of Nicaea. Further references to the Cloth of Edessa were made at this Council. Leo, the Lector of Constantinople, reported having seen it in Edessa (this has been referred to in the previous Chapter). The Council also made specific mention of the image "not made with human hands" that was sent to Abgar. It went on to say in its condemnation of iconoclasm that "one can and must be free to use images of our Lord and our God, in mosaics, paintings, etc" and that "the icon must be an image that bears a very close resemblance to its prototype", possibly a further reference to the Cloth of Edessa[2]. As we have seen, the Cloth of Edessa must have had great significance for the opponents of iconoclasm.

A further outbreak of iconoclasm started in the early ninth century under the Emperor Leo V the Armenian. This continued until 842, when icons were finally restored to the churches of the Empire[3].

* * * * * * * *

The troubled period of iconoclasm was followed by a period of internal recovery for the Empire. There was a revival of the arts and further growth of Byzantine culture. In 867 the Imperial throne was seized by a Macedonian, Basil I, who founded a dynasty of strong Emperors that continued to rule into the eleventh century. Basil and his son, Leo VI, were responsible, among other benefits for the Empire, for the creation and publication of a vast legal code, the Basilics.

In due course the son of Leo VI came to the throne as Constantine VII Porphyrogenitus ("born to the purple"). Unfortunately he was a child at the time and the leading admiral of the Empire, Romanus Lecapenus, took the opportunity to usurp the throne. Unusually, however, he did not kill the child he had superceded, but brought him up in his own Court and married him to his daughter Helena. By the year 944 Romanus saw that his reign was coming to an end; he was an old man and his reign had been a time of continued stability in the Empire. The Moslem armies that had earlier posed a major threat were now more on the defensive. In the light of what he then did it seems that he felt that he should leave to his Empire a lasting legacy. He decided to obtain the Cloth of Edessa and bring it to Constantinople.

Although it was a time of stability and strength for the Empire itself there was still mutual antagonism between the Muslim Caliphate and the Christian Empire. Sending an army to Edessa could easily have been interpreted as a major act of aggression and a cause for renewed war. Romanus took the chance, counting on the fact that the Caliphate was in a weakened state at the time. He sent an army under one of his leading generals, John Cuercas, who fought a campaign into Moslem territories that had not seen a Christian army for three hundred years. On reaching Edessa Cuercas promised the Emir of Edessa permanent immunity from attack, the release of two hundred Muslim prisoners and payment of twelve thousand silver pieces, all for one thing only – the Cloth of Edessa. Although the Emir sought and obtained permission from his superiors in Baghdad to hand over the Cloth, the Christians of Edessa were not happy to see it go. After all, not only was it sacred to them, it was also the source of considerable income from pilgrims. The Edessan clergy made two attempts to pass off copies of the Cloth to the Byzantines instead of the genuine article.

An intermediary for the handing over of the Cloth had been appointed – Bishop Abraham of Samosata, a nearby town. Eventually Bishop Abraham was satisfied that he had the genuine Cloth in his hands, that he could pass on to the Byzantines. The transaction was described by a contemporary chronicler in the following terms:

"…the city was in great turmoil as they would not allow their most precious possession to be taken away, which brought them protection against harm. Eventually the Moslem emir managed, by persuading some, using force against others, and terrifying others with threats of death, to get the image handed over. Then there was a sudden outbreak of thunder and lightning with torrential rain, as if by design or arrangement, just as the image and the letter of Christ were about to leave the city of Edessa. Once more those who had clung to these things before were stirred again. They maintained that God was showing by these events that it was not in accordance with Divine will that these most holy objects should be transferred."[4]

Cuercas retired from Edessa in some disorder, but with the Cloth in his possession. On his return to Constantinople the Cloth was taken first to the Church of St Mary at Blachernae (from which it would disappear 260 years later), then to the Imperial Place and finally to the Church of Hagia Sophia, where Gregory Referendarius delivered his sermon.

It is one of the ironies of history that this forced seizure of the Cloth probably guaranteed its survival. At the time of the First Crusade at the end of the eleventh century Edessa was captured by Christian forces and became an important part of the Crusader presence in the Middle East. In 1147 it fell once again, this time into the hands of the Saracens under Nur ed-Din. The native Christians were massacred to a man and their wives and children sold into slavery. The city was destroyed and left desolate. It has never recovered[5]. If the Cloth had remained in Edessa it would undoubtedly have been destroyed along with the churches and all other traces of a Christian presence there.

* * * * * * * * *

By the beginning of the thirteenth century Constantinople had the richest collection of relics in the Christian world. Some of them were described by the French crusader Robert de Clari, who described the interior of the Church of the Mother of God of the Pharos in the following terms:

"Within the Chapel were found many rich relics. There were two pieces of the True Cross as large as a man's leg and as long as half a *toise* (about three and a half feet); there was also the iron of the lance with which our Lord's side was pierced, and two of the nails which were driven through his hands and his feet. There was a crystal phial with some of his blood; there was the tunic he wore, which was taken from him when they led him to Mount Calvary; there was the blessed Crown with which he was crowned, and which was made of reed with thorns as sharp as the points of daggers...."[6]

One of the best-known of all the relics in Constantinople was known as the Mandylion. This has been identified in recent years with the Cloth of Edessa, but there are good reasons to believe that the Mandylion and the Cloth were two separate articles, which have become confused with each other over the long period of time that has passed since the thirteenth century.

The historian Sir Steven Runciman noted that the Mandylion could not be a shroud:

"The Image of Edessa was always described by the Byzantines as a *mandelion*, a kerchief, which is quite different from a *sindon* (or shroud)."[7]

Indeed the word "Mandylion" is derived from the Arabic word *mandil*, meaning a veil or handkerchief. This word in turn is derived from the Latin *mantile,* meaning a mantle. It could not therefore refer to a cloth the size of a shroud. In any event there is no suggestion in the term "Mandylion" of any association with burial procedures or cloths[8]. The appropriate word for a large burial cloth would be, as noted by Sir Steven Runciman, *sindon* or some derivative of this word.

One possible reason for the Cloth of Edessa being described as a *mandelion* rather than a *sindon* could have been that it was always kept folded. We have already seen how the term *tetradiplon*, or "doubled in four", was applied to the Cloth of Edessa. There is a further reference to the appearance of the Cloth in the Official History of the Cloth of Edessa, written in 945. It was said to have been fastened to a board by King Abgar of Edessa and embellished with "the gold which is now to be seen"[9]. Ian Wilson, who proposed a detailed hypothesis in his book *The Turin Shroud* identifying the Shroud with the Mandylion in Constantinople, showed quite clearly in diagrams in this book that when the Shroud is folded in a "doubled in four" manner the head appears in a disembodied fashion, and the rest of the body cannot be seen[10]. For this reason, he suggested, the full-length Cloth of Edessa was considered as no more than a kerchief or veil. This hypothesis, however, depends on it having been unknown in Constantinople that the Cloth of Edessa was a full body-length cloth. In fact, as we have seen, this was not the case. There were a number of references to the full length of the Cloth, and its full body image, while it was in Edessa and, notably, in the sermon of Gregory Referandarius on its arrival in Constantinople. The conclusion must be reached that while perhaps the Cloth of Edessa was generally kept in the "doubled in four" manner, it was also shown from time to time in its full form.

There is no doubt that there was a revered cloth known as the Mandylion in Constantinople, but there is also strong evidence that this was a different cloth to the Cloth of Edessa. The French crusader Robert de Clari, who was an eye-witness, attests to the fact that there were both a small image-bearing cloth and an image-bearing shroud in Constantinople, as two separate items. He described seeing the Shroud in the following terms:

"…there was another of the churches which they called My Lady St Mary of Blachernae, where was kept the *sydoine* (shroud) in which Our Lord had been wrapped, which stood up straight every Friday so that the features of Our Lord could be plainly seen there…"[11]

Elsewhere he describes seeing another cloth, in another place, with an image of the face of Christ on it:

"There were two rich vessels of gold hanging in the midst of the chapel by two heavy silver chains. In one of these vessels was a tile and in the other a cloth. And we shall tell you where these relics came from. There was once a holy man in Constantinople. It happened that this holy man was covering the house of a widow with a tile for the love of God. And as he was covering it, Our Lord appeared to him and said to him (now this good man had a cloth wrapped about him) 'Give me that cloth' said Our Lord. And the good man gave it to him, and our Lord enveloped his face with it so that his features were imprinted on it. And then he handed it back to him... but before he carried it away, after God had given him back his cloth, the good man took and hid it under a tile until vespers. At vespers... he took the cloth, and as he lifted up the tile, he saw the image imprinted on the tile just as it was on the cloth..."[12]

The shroud and the cloth are clearly described as two separate items, in two separate places. The shroud was in a chapel in the Blachernae Palace; the cloth and the tile were in the Church of the Mother of God of the Pharos in the Bucoleon Palace. These two places were separated by some distance. The tile described by de Clari is clearly the Keramion or Holy Brick which, according to tradition, had come into contact with the Mandylion and had thus been imprinted with the face of Christ. This had been brought to Constantinople from Hieropolis in Syria in 966[13]. The Keramion and the Mandylion are linked in tradition and it is not unreasonable to conclude that the cloth kept in the Bucoleon Palace with the Keramion was, in fact, the Mandylion. Therefore the Shroud, or Cloth of Edessa, which was in the Blachernae Palace at the time, could not have been the Mandylion.

The link between the Mandylion and the Keramion goes back several centuries. The two are closely linked in legends from Georgia, where the Keramion is said to have been taken in the sixth century by Anton of Martqopi, one of the fathers of the Syrian church.[14] According to a legend dating from the 12th – 13th centuries, the Mandylion was originally taken from Hieropolis in Syria to Georgia by the apostle Andrew. Other legends suggest that the Mandylion and the Keramion were taken to Georgia at the same time, by Anton of Martqopi.[15]

From this it may be possible to draw the conclusion that the Mandylion was one of the "images made without hands" which were in fact early copies of the Cloth of Edessa, referred to in the previous Chapter, which had started to circulate in the area around Edessa, including Hieropolis. Somehow both the Mandylion and its copy on a tile, the Keramion, had either traveled to Georgia or been associated with Georgia in some way on their journey from the Edessa/Hieropolis region to Constantinople.

It is also worth noting that the term "Mandylion" is not used in any contemporary reference to the arrival of the Cloth of Edessa in Constantinople. Its earliest known usage is in about 990 when it was used in the biography of an ascetic, Paul of Mount Latros, who had a vision of a cloth imprinted with the face of Christ[16].

It was clearly known at the time that the Cloth of Edessa, despite being kept folded, was a full-length cloth rather than just a face-cloth. The previous Chapter referred to the Greek theologian St John of Damascus describing the Cloth of Edessa as a *himation*, a long garment worn by the ancient Greeks. In the tenth century Leo the Deacon spoke of the image as being on a *peplos*, another term for a full-length robe[74]. Finally, in his sermon in 944, Gregory Referendarius made a clear reference to the wound in the side of the figure on the Cloth, when he said:

"For these are the beauties that have made up the true imprint of Christ, since after the drops fell, it was embellished by drops from his own side. Both are highly instructive – blood and water there, here sweat and image....The source of living water can be seen and it gives us water, showing us that the origin of the image made by sweat is in fact of the same nature as the origin of that which makes the liquid flow from the side.[18]"

Such a reference could only have been made if Gregory had been aware of the wound in the side of the figure and of bloodstains in the area of that wound, and hence had known that the cloth was a full-length cloth rather than just a face-cloth.

Arriving at the conclusion that the Mandylion and the Cloth of Edessa were two separate objects has great significance when it comes to trying to trace the movements of the Cloth of Edessa after the sack of Constantinople in 1204. This will be covered in detail in Chapter 9.

The Mandylion is not to be confused with another image-bearing cloth which may also have been an early copy of the Cloth of Edessa – the Veil of Veronica. The legend associated with this cloth is as follows. Veronica was in her house when Christ was led past on his way to Calvary. Moved by the sight of him struggling beneath the cross, she took her veil and wiped the sweat and blood off his face. When she returned inside her house, she found that he had left the imprint of his features upon it[19]. The Veil of

From Edessa to Constantinople

Veronica was in Rome at least from the early eleventh century – in 1011 Pope Sergius IV consecrated an altar to the Veil in St Peter's, quite possibly to mark its arrival in Rome. In 1191 it was recorded that Pope Celestine III had shown it to the visiting King of France, Philip Augustus[20]. It has been suggested that this cloth was destroyed in 1527 when the army of the Holy Roman Emperor Charles V sacked Rome[21]. However, the historian Ian Wilson has produced evidence that the Veronica was still in existence in 1617[22]. A relic known as the Veronica continues to be preserved in St Peter's in Rome, but an eye-witness description of it in the 1950s describes it as containing "a head-size patch of colour.....Even with the best imagination, you could not make any face or features out of them.[23]" If this is indeed the original Veronica, it has not stood up to the ravages of time.

* * * * * * * * *

While the Cloth of Edessa was in Constantinople a number of representations of it were made. Possibly the most significant of these is contained in a Hungarian manuscript known as the Pray Manuscript. This manuscript has great importance to Hungarians as it is the oldest surviving text in their language, having been created at the Boldva Benedictine monastery in about 1192. It is also of major importance in the history of the Shroud.

The Pray Manuscript has four pages of coloured drawings; the third of these pages shows in its upper section Joseph of Arimathea and Nicodemus preparing the dead body of Jesus for burial. In the lower section there is a depiction of the three Marys visiting the tomb which is being guarded by an angel.

In the upper section the body of Jesus is shown in a pose remarkably similar to that of the man on the Turin Shroud. The body is also totally naked – a very unusual representation for the time. There are further remarkable similarities between the drawing in the Pray Manuscript and the Turin Shroud. The drawing shows all four fingers of Jesus's hands, but no thumbs. This is exactly what is depicted on the man on the Turin Shroud. (There are specific medical reasons why the thumbs do not appear on the man in the Shroud – these will be discussed in a later Chapter.)

In addition the drawing shows a single bloodstain on Jesus's forehead, just above the right eye. This is located in exactly the same position as a distinctive '3'-shaped stain on the forehead of the man on the Shroud. The Pray Manuscript drawing also shows the nail wound in Jesus's right hand to be through the wrist rather than through the palm, another very unusual representation for the time which is matched on the man on the Shroud.

Most remarkable of all, in the lower section of the page there is a shroud, partly rolled up, on the lid of the sarcophagus representing Jesus's tomb. A close study of this shroud shows a set of four tiny holes, three in a line and one off-set, corresponding precisely to the "poker holes" on the Shroud of Turin. There is a further set of holes in exactly the same pattern on the sarcophagus lid. These features in the Pray Manuscript were studied by a French scientist, Professor Jerome Lejeune, who concluded:

"Such precise details are not to be found on any other known (Christ) image – except the Shroud that is in Turin. One is therefore forced to conclude that the artist of the Pray Manuscript had before his eyes… some model which possessed all the characteristics of the Shroud which is in Turin."[24]

In addition to the features on the shroud shown in the Pray Manuscript, the tomb slab itself is strangely decorated with an unusually flamboyant Greek key motif. It has been argued that the artist was naively representing the herringbone twill of the Shroud of Turin[25]. This herringbone weave is one of the specific features of the Shroud. It has been remarked upon by scientists during studies in the late twentieth century and is clearly shown on a medallion dating to the mid-fourteenth century showing the Shroud together with the coats-of-arms of Geoffrey de Charny and his wife Jeanne de Vergy. (See Chapter 4.)

This can only be described as striking evidence that the Shroud of Turin was in existence, and known, prior to the start of the thirteenth century. It is only a small further step to identify the Shroud of Turin with the Cloth of Edessa.

Another possible representation of the Cloth of Edessa is shown in a type of painting which began in the late eleventh and twelfth centuries. This type of painting is known as Extreme Humility or Utmost Humiliation, or, alternatively in the West, as the Man of Pity or Man of Sorrows. In some of these representations the man is shown rising out of a box that is clearly too small to contain the rest of his body, but which could contain a folded cloth[26]. In other representations the man is shown rising from a tomb. The Byzantine scholar Hans Belting has expressed his opinion that the Man of Sorrows icon is a visual expression of its ritual function[27].

This is a reference to the Easter rituals of Edessa, prior to the transfer of the Cloth of Edessa to Constantinople in 944. In these rituals the entire body of Jesus shown on the Cloth was never shown to the faithful close up[28]. Rather it was revealed in stages, as described by Pope Stephen III in the eighth century:

"At the first hour of the day (Christ appeared) as a child, at the third as a boy, at the sixth as an adolescent and at the ninth hour visible in his full manhood, in which the Son of God went to his passion when he bore for our sins the suffering of the Cross."[29]

This description fits with texts by Nicholas Mesarites and Robert de Clari, both of whom referred to the shroud in Constantinople as gradually rising or standing up in the manner depicted in Man of Sorrows icons[30].

One of the best known and earliest examples of Man of Sorrows iconography is the Image of Pity in the Church of Santa Croce in Gerusalemme in Rome. This is a mosaic by an Italian master but in the Byzantine style, dating to the thirteenth century. It shows Christ's body down to the waist with the hands crossed, as in the Shroud, but over the abdomen rather than the pelvic area. The bleeding wound in the man's side is well in sight[31]. These features of crossed hands and a bleeding wound in the side are generally typical of Man of Sorrows art.

Another art form that developed in the 11th – 12th centuries and which can possibly be related to the Cloth of Edessa was that of *epitaphioi threnoi* ("funeral lamentation" cloths). Until this time Jesus had been represented on embroidered cloths in a placid way. However an evolution then occurred by which Jesus came to be represented dead, sometimes life-sized, upon his shroud. The earliest of these *threnos* representations appeared in Serbia, Macedonia and, a bit later, in modern-day Romania. It is unlikely that the artists would have seen the original Cloth in Constantinople. However Hans Belting, the Byzantine scholar already referred to, has suggested that there is a clear link between the appearance of this art form and what he described as "...what was believed to be the authentic relic of the Holy Shroud preserved in the chapel of the Palace (in Constantinople)...[32]"

Although it has been suggested that the Cloth of Edessa was never publicly exhibited for much of the time it was in Constantinople[33], details of the image were clearly known and represented in different forms of art from at least the late twelfth century onwards.

* * * * * * * * *

The Cloth of Edessa was mostly kept with the other relics of the Passion in the Imperial Bucoleon Palace. There were times, however, when it was recorded as being in the Blachernae Palace. In 1147 the French King Louis VII visited the Emperor Manuel Comnenus in Constantinople, where he venerated the Cloth of Edessa in the Blachernae Palace[34]. In 1171 it was back in the Pharos Chapel of the Bucoleon Palace, where it was shown to King Amalric I of Jerusalem. On this occasion specific reference was again made to a *sindon,* or shroud[35].

In 1201 it was still in the Bucoleon Palace, in the Pharos Chapel. The treasurer of the Pharos Chapel, Nicholas Mesarites, in that year warned the enemies of the Byzantine Emperor not to attack the palace:

"In this chapel, Christ rises again, and the Shroud with the burial linens is the clear proof....They still smell of myrrh and are indestructible since they once enshrouded the dead body, anointed and naked, of the Almighty after his Passion."[36]

Two years later, however, it was reported as being in the Church of St Mary in Blachernae by Robert de Clari. The reason for its move was, without doubt, the arrival at Constantinople of the army of the Fourth Crusade. The Cloth of Edessa, known for its powers as a palladium, or protection against enemies, had been placed on public view in order that its power would protect Constantinople from this latest and extremely powerful enemy.

From Edessa to Constantinople

The Fourth Crusade was one of the seminal events in world history, resulting as it did in the capture and sack of the centre of Greek culture and civilization, Constantinople, by Crusaders from northern and western Europe. Although the Latin Empire established at Constantinople by the Crusaders did not last long and the Greek Byzantine Empire was re-established, the latter was fatally weakened. The long-term result was the fall of the Byzantine Empire and the firm establishment of an expansionist Moslem power in south-eastern Europe.

Constantinople was not the original objective of the Fourth Crusade, although some writers have suggested otherwise[37]. The original objective of the Fourth Crusade was to succeed where the Third Crusade had failed – in the recapture of Jerusalem from the Saracens, who had captured it after the Crusader disaster at the Horns of Hattin in 1187. In 1199 Count Tibald of Champagne, a nephew of Richard Coeur-de-Lion, the King of England, and a relative of other leading Crusaders, met with some friends for a tournament at his castle. At the end of the tournament they vowed to take the Cross and sent off a messenger to inform the Pope.

Pope Innocent III had come to the Papal throne in 1198 and had already expressed his desire for a new Crusade. In 1199 he sought a detailed report on the situation in Palestine. He welcomed the decision of Count Tibald and his friends[38].

As was usual at the time, nothing happened quickly. Based on the opinions of King Richard I of England it was decided that Egypt rather than Jerusalem should be the Crusaders' objective, as this was perceived to be the vulnerable point in the Saracen Empire. In 1201 Tibald died suddenly and leadership of the Crusade was taken over by Boniface of Montferrat. The House of Montferrat also had many connections with the East and Boniface had been the uncle of King Baldwin V of Jerusalem. Boniface then went off to Germany to spend some time with his friend Philip of Swabia, a connection that was to lead to the disaster at Constantinople. Philip of Swabia was a son of the Holy Roman Emperor Frederick Barbarossa and a contender for the throne of the Holy Roman Empire. He married Irene Angelina, the daughter of the Byzantine Emperor Isaac II Angelus, and this led him to take an interest in the dynastic affairs of Constantinople.

Frederick Barbarossa and his family harboured a deep hatred for the Byzantine Empire, which they saw as a rival to their own Holy Roman Empire in Germany. Byzantine traditions outrivaled those of Germany and there was the on-going schism between Latin Christianity in the West and Greek Christianity based in Constantinople. Moreover the Byzantine Empire lay in the way of German ambitions to establish a Mediterranean dominion. Philip of Swabia's brother, Emperor Henry VI, had been planning a campaign in the East, which could well have involved the conquest of Constantinople, when he died suddenly at an early age[39].

Philip's father-in-law, Isaac II of Constantinople, was deposed in a palace coup in 1195 by his brother, who became Alexius III. Isaac had not been a very successful emperor, but neither was Alexius III. Isaac's son, another Alexius, had been imprisoned with him, but he escaped in 1201 and made his way to the court of his brother-in-law Philip in Germany, where he was introduced to, among others, Boniface of Montferrat.

Meanwhile the Crusaders had been negotiating with Venice for a fleet to transport them to Egypt. These negotiations and preparations were prolonged but eventually the Venetians provided the required fleet, provided the Crusaders first recaptured for Venice a city in Dalmatia, called Zara, which the Venetians had recently lost to the King of Hungary. The Crusaders obliged by capturing and pillaging the city in November 1202, having sailed to the Dalmatian coast in the Venetian fleet, ostensibly still on the way to Egypt. Pope Innocent III was appalled that a Crusade should have been used to attack another Christian kingdom, but there was little he could do except excommunicate the Venetians.

Now Alexius saw an opportunity. He arranged for Philip of Swabia to send a message to Boniface of Montferrat at Zara proposing that, if the Crusade would proceed to Constantinople and place him (Alexius) on the imperial throne there, he would pay them money still owed to the Venetians and assist them in their conquest of Egypt. He also offered the submission to Rome of the Church in Constantinople. This thoroughly suited the Venetians, who wanted to strengthen their trading position against Constantinople and who had their own reasons for delaying an attack on Egypt. The Pope was most unhappy about this, as he saw no good coming from a scheme involving Venice and the friends of Philip of Swabia. However, he restricted himself to issuing an order that no more

Christians were to be attacked unless they were actively hindering the Holy War[40].

Alexius then joined the Crusaders at Zara and they all moved on to Constantinople, where they arrived in June 1203. Alexius III was not prepared for this threat. However Constantinople was defended for a while against the Crusader army before Alexius III fled. In the city Isaac II, who had been blinded when he was deposed, was brought out from his captivity and restored to his throne. The Crusaders insisted on Alexius being crowned as his father's co-Emperor and abiding by the promises that he had made them. Alexius thus became Alexius IV. He proved as incompetent as his father and uncle had been. His efforts to make the Greek clergy admit the supremacy of Rome met with resistance and he was unable to find either the men or the money he had promised the Crusaders. The situation deteriorated, with mutual resentment and antagonism between the Byzantines and the Crusaders[41].

Alexius III had had two daughters. One of them, Eudocia, had married a man called Alexius Murzuphlus. In a popular uprising in early 1204 Alexius IV was overthrown and murdered. His father Isaac II died a few days later and Alexius Murzuphlus became Emperor as Alexius V. This was a direct challenge to the Crusaders. The Venetians, under their Doge Dandolo, urged the storming of Constantinople and the installation of a Western Emperor. One nominee was Philip of Swabia, who would unite the two Empires, but he was far away in Germany. Another was Boniface of Monteferrat, but the Venetians disapproved of him. Finally it was agreed to appoint a panel to elect a Latin Emperor as soon as the city was taken.

The attack took place in April 1204. A breach was made in the city walls and the Crusaders and Venetians poured into the city. Alexius V fled, as did other members of the imperial family. Once the Doge of Venice and the Crusaders were established in the Great Palace they told their soldiers they could spend the next three days in pillage. Constantinople was put to the sack[42].

At the time Constantinople was the greatest and richest city in the world. Great wealth was seized by the Crusaders. Many treasures, including religious objects, paintings, relics and books were destroyed.

The Cloth of Edessa disappeared in the chaos and confusion.

Chapter 4 –
Geoffrey de Charny and the Shroud of Lirey

In July 1352, in the town of Saint-Omer in northern France, a man was led out to die in front of a large crowd that had gathered for the spectacle. As was not uncommon for that period his death was to be gruesome and shocking – an act of both revenge and deterrence. He was tortured with hot irons and then dismembered with a meat axe[1]. His remains were displayed on the town gates[2]. His name was Aimeric of Pavia and his crime was that he had double-crossed one of the leading soldiers of France, Geoffrey de Charny.

In 1347, following the disastrous defeat of the French army by the English at Crecy the year before, the English had captured Calais on the English Channel after a lengthy siege. The capture of the town was followed by a truce between the two warring armies. The truce lasted until 1349, when hostilities resumed, at which time Geoffrey de Charny devised a plan to recapture the town. His plan depended upon the co-operation of an Italian mercenary, the unfortunate Aimeric, who was at the time the galley-master in Calais of the English King, Edward III. He agreed to admit de Charny's troops into the citadel at night, in return for an enormous bribe, estimated at 20 000 *ecus*. However, he reported the plot to the King, who promptly crossed over the Channel from England with a small army. De Charny had no idea the King was in the town. With an army of some 5 500 men he approached Calais early in the morning of New Year's Day 1350. Aimeric came out and received the first part of his bribe; he then lured the leading companies of the French into the town, where they were promptly set upon and captured by the English. De Charny was then attacked by English forces led by both the King and the Prince of Wales. De Charny was badly wounded in the ensuing fight, which became a rout of the French, and was captured and taken to London. He eventually returned to France in 1351, after a ransom suitable to his position had been paid. Aimeric kept his money and went on a pilgrimage to Rome[3].

The following year de Charny besieged a castle in the town of Guines, near Calais. Eventually he had to abandon the siege, but he had learned that Aimeric had returned to France and was commanding a small English outpost close to Calais. De Charny rapidly captured this outpost and dealt with Aimeric in a manner he obviously felt appropriate[4].

De Charny was no more savage than any other soldier of his time. Indeed he had earned himself a reputation as a paladin and example of chivalry. He wrote books on chivalry and one of his works, *Le Livre de Chevalerie*, has long been recognized as one of the pioneer works on the subject. He was one of the leading men of France and was tasked by his King on at least one occasion with delicate political negotiations[5]. He had borne the French royal standard, the *Oriflamme*, in an army that had failed to relieve Calais in 1347[6]. He was to die bearing the *Oriflamme* at the Battle of Poitiers in 1356[7].

De Charny had been a captive of the English once before. He was captured at the Battle of Morlaix in Brittany in 1342, the first pitched battle of the Hundred Years' War, which had been another rout for the French army and where he had commanded a cavalry unit. He may have been released very quickly on this occasion, as he was back with a French army near Vannes in Brittany towards the end of that year[8]. He was knighted in 1343[9].

In 1346 de Charny took part in a campaign to Smyrna in Asia Minor, led by the ruler of the Dauphine in southern France, Humbert II de Viennois. This was described as a Crusade as it was a campaign against the Turks. It achieved very little and de Charny clearly returned to France with the minimum of delay, as he is known to have been at the siege of Aiguillon on the River Garonne in south-west France in August that year. Some writers have speculated that he might have obtained the Shroud in Greece at this time, but this can be considered most unlikely[10]. These speculations are just that – speculations, with no documentary or other supporting evidence.

Although not a member of the high nobility, Geoffrey de Charny came of a distinguished family. He could claim among his ancestors Pierre de Courtney, who had reigned as Latin Emperor of Constantinople in the early thirteenth century. This gave him a distant connection to the Valois Kings of France and could partly explain his rise to political prominence under Philip VI and John II. Philip's father, Charles de Valois, had married as his second wife Catherine de Courtney, another descendant of the Latin Emperors of Constantinople and the titular heiress to that throne. De Charny's mother was the daughter of Jean de Joinville, a famous chronicler of the Crusades who had been a close companion of King Louis IX[11].

De Charny first entered recorded history in 1337, the year in which the Hundred Years' War started. He was described at the time as a *bachelier*, a knight who has not yet acquired a fief by marriage or inheritance. In that year he took part in campaigns in Languedoc, in the south of France, and Guyenne under the High Constable of France, Raoul de Brienne. In 1340 he was recorded as being among "the flower of chivalry" who defended Tournai against the army of King Edward III. In 1341 he was in Angers with the Duke of Normandy, the future King John II of France, on a campaign against the English in Brittany[12].

De Charny made his name as a soldier in campaigns all over France. But not only was he a soldier. He was also a deeply religious man. In 1343 he petitioned King Philip VI for financial help in building a chapel in his home town of Lirey in Champagne[13]. The King responded favourably. In 1349 he confirmed a donation of land yielding an income that would pay the salaries and expenses of the canons who would take charge of the Lirey church. That same year de Charny wrote to the Pope, Clement VI, in Avignon, telling him of his intention to build a church in Lirey. Some years later, in 1353, he obtained a further rent from Philip VI's successor, John II, for the foundation of his church, and the church was formally founded at the beginning of July that year[14].

De Charny's deep religious belief can be seen in his own writing. He concluded his *Le Livre de Chevalerie* with the following words:

"For you should be certain of and hold firmly to the belief that you have no other course of action to take except to remember that if you love God, God will love you. Serve Him well: He will reward you for it. Fear

Him: He will make you feel secure. Honour Him: He will honour you. Ask of Him and you will receive much from Him. Pray to Him for mercy. He will pardon you. Call on Him when you are in danger: He will save you from it. Turn to Him when you are afraid, and He will protect you. Pray to Him for comfort, and He will comfort you. Believe totally in Him and He will bring you to salvation in His glorious company...."[15]

* * * * * * * * *

The first decades of the Hundred Years' War devastated France militarily, politically, economically and socially. The war had started in 1337 with fighting in what is now the Netherlands and Belgium as well as in Gascony. France's first major defeat had come in 1340 with the destruction of her fleet at the Battle of Sluys. Of 213 French ships present at the battle, the English captured 190. The French dead totaled between 16 000 and 18 000 and included both French admirals. It was a naval catastrophe unequalled until modern times[16]. This was followed by campaigns in Brittany and Gascony which led to further French defeats, including Morlaix. In 1346 Edward III landed in France with an English army and marched on Paris, having plundered the town of Caen in Normandy. Unable to enter Paris the English King marched north and eventually met the French army in battle at Crecy. Crecy was a disaster for the French. The English longbow showed its clear superiority to the crossbow and over 1 540 French knights and squires were killed together with an unknown number of infantrymen[17]. The operations of English armies in Brittany, Gascony and northern France caused enormous disruption to civil life in those regions, with consequent social and economic effects.

In addition to military defeat and the ravages of English armies, France had also experienced the horror of the Black Death in 1348. The Black Death combined three different forms of plague – bubonic, pneumonic and septicaemic – in a particularly lethal combination. Between sixty and ninety percent of victims of bubonic plague could expect to die, and when death occurred it was up to a week after infection. Recovery from pneumonic plague was virtually unknown, and the disease resulted in death in less than two days. Septicaemic plague kills within hours. There were cases reported during the Black Death where a man would go to bed in the best of health and be dead by morning[18].

The Black Death probably came to France at the beginning of 1348, brought to Marseilles by Genoese galleys that may have come from the Crimea. Within a month up to fifty-six thousand people had died in Marseilles alone and the plague was spreading rapidly through France. By June 1348 it had reached Paris and did not die out there until the winter of the following year. The death toll in Paris was reported as being fifty thousand out of a total population of some two hundred thousand. In the Papal city of Avignon more than half the population is said to have died. Mortality was probably less in the countryside than in the towns although some small communities seem to have lost half of their population. The plague died out in France by 1350[19], although there were several more, but lesser, recurrences of it later in the fourteenth century.

Frenchmen of the time must have thought that God had abandoned them. As one writer put it, writing about northern France in particular:

"To the luckless villagers, whose few possessions had alternately been looted by French or English soldiery, the apparition of the plague seemed merely the culminating phase in a process designed by God to end in their total destruction[20]."

It is quite understandable that a patriotic and religious Frenchman of high rank would seek the intervention of God on behalf of his country. Geoffrey de Charny had started his plans for his church in Lirey with his petition to Philip VI in 1343. Nothing seems to have been done for some years and it would not be surprising if the money promised by the French King had never materialized. At that time de Charny had not achieved the importance he gained in the last years of Philip VI's reign, when he was a member of the King's council, and under his successor John II. De Charny's church eventually came into being in 1353. Somewhere and somehow he came into possession of the Shroud. Obviously he knew where he obtained it from. Bearing in mind the sort of man he was, it is most unlikely that he would have had any truck with an obvious fraud. He must have had good reason to believe that the Shroud was the true burial cloth of Christ. At the very least it must have been for him an item of great religious significance because he exhibited it in the same church that he had founded and that had been the subject of correspondence between him and two kings and one pope.

The Shroud seems to have been first exhibited in Lirey in about 1355. It attracted large crowds of pilgrims and special souvenir medallions were struck for these pilgrims. These medallions, one of which still survives in the Cluny Museum in Paris, showed a representation of the Shroud and the arms of de Charny and his wife, Jeanne de Vergy[21]. This is particularly significant, as it may be an indication that Geoffrey de Charny was still alive at the time of the exhibition of the Shroud, and hence this must be dated before his death at the Battle of Poitiers in 1356[22]. On the other hand documents dated 1354 list various relics associated with the foundation of the church in Lirey, but make no reference to the Shroud[23].

It is not fanciful to suggest that the exhibition of the Shroud was de Charny's way of seeking God's intervention for his country. He had founded a church and he had an important religious artifact which he exhibited to attract pilgrims to pray there. To a man of his time and nature these would have been deeply religious acts with deep prayerful significance. He may well have been influenced by the exhibition of the Veil of Veronica in Rome during the Holy Year in 1350. The exhibition of the Veil of Veronica at this time was widely known throughout Europe and it is inconceivable that de Charny would have had no knowledge of it[24].

To suggest that Geoffrey de Charny would wilfully have used a forgery of no value to try and extract money from gullible pilgrims is to totally misunderstand the type of man he was and the time in which he lived. He was so convinced of the Shroud's religious importance that he associated it with his own coat of arms – a significant act for a man to whom chivalry was a way of life.

1355 was a difficult year for France and one which certainly merited God's intervention. After several years of relative military calm, campaigns began to get underway again. The main pre-occupation of the King of France and his ministers at the time was to get rid of the threat to France posed by Charles of Navarre[25], a relative and son-in-law of the King who had played a murky and underhand political role for a number of years. To do this the King appointed his heir, the Dauphin Charles, as his Lieutenant in Normandy to seize Charles of Navarre's possessions in that province. Part of the manoeuvres between the King of France and Charles of Navarre involved a landing in Normandy by an English army under the Duke of Lancaster in alliance with Charles. Charles changed sides once again, abandoning his alliance with the English, and the landing in Normandy had to be cancelled. Later in the year an English army under Edward III carried out a campaign in Picardy. The major threat to France, although it was not apparent at the time, came from the Prince of Wales in Gascony. He led an army into Languedoc and achieved a number of successes. After that campaign, in mid-1356, he turned north and marched towards central France. At the same time an English army under Henry of Lancaster finally landed in Normandy.

The Shroud may have been continuously exhibited in Lirey at this time. In May 1356 Henri of Poitiers, the Bishop of Troyes, formally ratified Geoffrey de Charny's letters instituting the church of Lirey. Bishop Henri warmly praised de Charny, saying "we praise, ratify and approve the said letters in all their parts". He went on to speak of "ourselves wishing to develop as much as possible a cult of this nature" and referred to de Charny's "sentiments of devotion, which he has hitherto manifested for the divine cult, and which he manifests ever more daily"[26]. If this is a reference to de Charny's exposition of the Shroud at his church, it clearly indicates firm approval for it. Indeed, the following year, after de Charny's death at Poitiers, twelve bishops at the Papal Court in Avignon granted indulgences to all who visited the church of St Mary at Lirey and its relics[27].

At this time the Shroud was clearly recognized by Church authorities as being of religious importance. How de Charny came into possession of it is not known, but he obviously would have known where he obtained it from and what its immediate history was. He would clearly have been in a position to judge its religious worth. His judgement of this worth seems to have been accepted and shared by leading churchmen at the time.

In September 1356 the armies of the Prince of Wales and the King of France met at Poitiers. The outcome was catastrophic for France. While Geoffrey de Charny died a hero's death defending the royal standard, the *oriflamme*, and the King in a last stand, the French army was decisively defeated and the King of France captured by the English. France descended into anarchy. In the absence of the King in captivity in London, Charles of Navarre was free to cause havoc. He instigated revolution in Paris which continued through 1357 and 1358. In June 1357 there was a peasants' revolt, known as the Jacquerie, which only lasted a few weeks but which spread into the region of Champagne, in which Lirey lies. This was directed primarily against the nobility in manors and castles, but there was a degree of indiscriminate looting. The period 1357 – 1359 also saw the rise of the Companies, hordes of soldiers under independent command that spread out through France, occupying castles and manors and extorting whatever they could from the unfortunate population[28].

Bands from Gascony had moved west into Auvergne as early as 1355, but these were relatively small groups of men, a hundred or two hundred strong at most. In 1357 the "Great Company" was formed by one Arnaud de Cervole, who was also known as the Archpriest. This consisted of a succession of large temporary armies, including unemployed foot soldiers and archers, professional criminals from the towns, penniless gentry and a handful of knights, that rampaged across southern France looting and destroying. Other companies of English and Navarrese soldiers operated in the region of Paris and Amiens. Yet another notorious leader was Robert Knolles, an Englishman who raided into the Auxerre region in 1358 and 1359. In early 1359 Knolles' company established a base only twenty miles from the cathedral town of Troyes and then marched on Troyes itself. Troyes was defended energetically, one of the leaders in this endeavour being the Bishop, Henri of Poitiers. Knolles failed to capture Troyes but his bands remained in the area for nearly five months. In March he captured Auxerre itself. Jonathan Sumption in *The Hundred Years War – Vol II: Trial by Fire*, gives a vivid description of how these mercenaries operated once they had seized Auxerre:

"This was to be a careful appropriation of the city's wealth by disciplined professional thieves whose procedures were calculated to maximize their profits. Hardly anyone was killed. Soldiers posted at the gates stopped the inhabitants from leaving so that they could be assessed for ransom. The floors were dug up and secret caches hacked out of the walls in which they had been hidden. In the cathedral the great silver lamps which hung before the high altar were taken down and the treasury was emptied. The portable booty of a city of twenty churches must have been prodigious[29]."

Lirey is only twelve miles from Troyes and hence was vulnerable to the depredations of the Companies. The Champagne region around Lirey had also been affected by the Jacquerie. The Shroud had been exhibited in the small wooden church in Lirey since about 1355 and appears to have become quite well known. Such a valuable religious artifact in a small and indefensible village would undoubtedly have been an attraction for the groups of bandits raiding throughout France in the late 1350s. It would not be at all surprising if the owner of the Shroud at that time, the widow of Geoffrey de Charny, stopped the expositions and hid the Shroud away to protect it. In any event, the expositions seem to have stopped at about that time and the Shroud once again left the historical record, for a short time at least.

* * * * * * * * *

There is continuing debate among historians as to the exact year in which the exhibition of the Shroud in Lirey started. 1355 is chosen as a date on the basis of a memorandum written in 1389 by Bishop Pierre d'Arcis of Troyes (discussed at length in the next Chapter), in which he describes the exhibition as having been stopped "thirty-four years ago or thereabouts"[30]. However there is no contemporary written record of the Shroud having been exhibited in Lirey during the lifetime of Geoffrey de Charny and church records of the time make no reference to any exhibition of the Shroud.

One possibility is that the letter of Henri of Poitiers, Bishop of Troyes, written on 28 May 1356, praising, ratifying and approving Geoffrey's foundation, was the granting of formal approval for the exhibition of the Shroud at Lirey. King John II ratified the foundation in July 1356. Geoffrey must have left Lirey for the last time in June 1356. He was entrusted with the *Oriflamme* on 30 June, which required him to be with the king, and he died on 19 September 1356[31]. Geoffrey may well have intended to exhibit the Shroud at his church in Lirey, for all the reasons detailed earlier in this Chapter, but time may have caught up with him and he may never have had the opportunity to organize it himself.

The major piece of supporting evidence for the exhibition of the Shroud in the 1350's, apart from Bishop d'Arcis's memorandum, is the pilgrim badge found in the River Seine. This contained the coat of arms of Geoffrey de Charny together with those of his wife, Jeanne de Vergy. The crucial question is this, why were Jeanne's arms on the badge?

One possible answer to this is that ownership of the Shroud passed in September 1356 to the infant Geoffrey II de Charny, who was under the guardianship of his mother. She was also an heir to her husband's possessions. In this situation the juxtaposition of the coats of arms of mother and son (not wife and husband) would have been quite normal. The indulgence granted in 1357 by the twelve bishops of the papal court in Avignon was not only for visitors to the church in Lirey but also for the salvation of Jeanne de Vergy. This was the first time she was named in church documents. Clearly she played a significant role in the exhibition of the Shroud and it could be argued that it was she who started the exhibitions at Lirey, after the death of her husband[32].

If the exhibition only started in late 1356, or even 1357, it could not have lasted long in view of the instability and increasing lawlessness in the country. However, the existence of the pilgrim badge would seem to indicate that the exhibition lasted some time and that the presence of the Shroud in Lirey became well known. Bishop d'Arcis also refers to this in his memorandum, when he says that the story that the Shroud was that of Christ was "put about not only in the kingdom of France, but, so to speak, throughout the world, so that from all parts people came together to view it."[33] Even making allowance for some degree of exaggeration on the part of the good bishop, this does seem to tell us that the exhibition had become widely known.

There is another possible reason for Jeanne's arms being shown on the pilgrim badge. It is possible that it was she, not her husband, who was the true owner of the Shroud and that it was the arms of her husband, as the founder of the church, which were associated with hers on the badge, rather than vice versa. This possibility will be explored in more detail in Chapter 9.

There can be no definite answer to the question of when the exhibition of the Shroud started in Lirey unless direct evidence of it is found in an archive one day. However, the fact that the exhibition lasted long enough to become widely known among the pilgrim community, and that it had to have been stopped by 1358 at the latest because of civil unrest, does suggest that it must have been started by Geoffrey de Charny himself, before his death at Poitiers.

Chapter 5 – The Later Years in Lirey

During the 1360s the Companies continued to trouble France, particularly in the south. There were a series of "Great Companies" that caused particular havoc. In 1362 a Great Company threatened Burgundy, immediately south of the Champagne region. At the Battle of Brigais that year the Great Company defeated an army of the King of France. Following this a group of the Great Company moved into Burgundy for a while, seizing vulnerable castles and living off the land[1].

King John II died in 1364 and was succeeded by his son, Charles V. King John had been a good man – indeed he was known as John the Good. However he had not been a particularly good king and he had shown himself a poor soldier and poor diplomat. His son was to prove himself very different. The following year he obtained the submission of Charles of Navarre and brought to a close the on-going civil war that had been waged by that individual. That same year the larger Companies moved into Spain to involve themselves in the dynastic disputes of Aragon and Castile. The smaller bands remaining in France had mostly disappeared by 1366 and some degree of order returned to rural France, after ten years of upheaval. This was once again rapidly disturbed following the victory of the Prince of Wales at Najera in Spain in 1367. The remnants of the Companies in Spain moved back to France and formed a new Great Company, which moved north during 1367 and raided into the Champagne region past Auxerre and Troyes, as far north as Epernay[2]. Again it would have been a time to lie low in Lirey. By 1369 this Great Company had been defeated and dispersed, but this year saw a resumption of the war between the French and the English. During the 1370s France recovered lost military ground under the military leadership of Bertrand du Guesclin. Charles V died in 1380, and for the next thirty years the Hundred Years' War between France and England was effectively suspended by constantly renewed truces[3].

* * * * * * * * *

Jeanne de Vergy was the second wife of Geoffrey de Charny. His first wife had died, possibly during the Black Death. Whether he had any children by his first wife is not known – certainly they did not survive into adulthood. The date of his marriage to Jeanne de Vergy is also not known, but their son, also called Geoffrey, must have been born in the early 1350s; in 1366 in a census return completed by his mother and her second husband he had been described as a squire who had not reached the age of majority[4]. The Shroud seems to have remained hidden during his childhood and early adulthood, due partly no doubt to the continuing unrest throughout France during most of the 1360s[5]. Like his father, Geoffrey II de Charny served in the armies of France. In 1382 at the Battle of Roosebeke, where the French defeated the Flemish, he attracted the attention of King Charles VI of France who praised his role in the battle[6].

By 1389 peace had been fully restored in most of rural France – certainly in the region of Champagne – and Geoffrey II de Charny was already approaching middle age. In this year, possibly at the suggestion of the then Dean of Lirey, Nicole Martin, Geoffrey II started exhibiting the Shroud again in the small church of Lirey. Again pilgrims streamed to the church to see it[7]. This attacted the ire of the Bishop of Troyes of the time, Pierre d'Arcis, a successor of Henri of Poitiers. Permission for the exhibition of the Shroud had been obtained directly from both the Pope and King Charles VI of France, bypassing the Bishop. Bishop Pierre wrote a lengthy memorandum, known to history as the "d"Arcis Memorandum", ostensibly to Pope Clement VII[8] at Avignon. In this memorandum he alleged that Bishop Henri had declared the Shroud to be a forgery. According to d'Arcis:

"The Lord Henri of Poitiers, of pious memory, then Bishop of Troyes, becoming aware of this (that the Shroud was a forgery), and urged by many prudent persons to take action, as indeed was his duty in the exercise of his ordinary jurisdiction, set himself earnestly to work to fathom the truth of this matter. For many theologians and other wise persons declared that this could not be the real shroud of our Lord, having the Saviour's likeness thus imprinted upon it, since the Holy Gospel made no mention of any such imprint while, if it had been true, it was quite unlikely that the holy Evangelists would have omitted to record it, or that the fact should have remained hidden until the present time. Eventually, after diligent inquiry and examination, he discovered the fraud and how said cloth had been cunningly painted, the truth being attested by the artist who had painted it; to wit, that it was a work of human skill and not miraculously wrought or bestowed. Accordingly, after taking mature counsel with wise theologians and men of the law, seeing that he neither ought nor could allow the matter to pass, he began to institute proceedings against the said Dean and his accomplices in order to root out this false persuasion. They, seeing their wickedness discovered, hid away the said cloth so that the Ordinary could not find it, and they kept it hidden afterwards for thirty-four years or thereabouts down to the present year (1389)[9]."

Bishop d'Arcy therefore appeared to be making the following allegations: the Shroud is a forgery, being no more than a painting; Bishop Henri obtained evidence of this from the very artist who had painted it; Bishop Henri had suppressed the expositions of the Shroud; and although he does not mention Geoffrey de Charny by name, he is implicitly accusing him of being a party to this deception.

The d'Arcis Memorandum is a highly controversial document. Firstly, no record of it has ever been found in any Apostolic Archive, casting doubt on whether it was ever actually received by the Pope. Secondly, two hand-written versions of the document exist in the Champagne vault 154 in the Manuscript Hall of the Paris National Library. These two versions are referenced as folio 137 and folio 138. Folio 137 has 66 lines of about 27 words each and folio 138 has 65 lines. Folio 138 appears to be the original draft of the document, containing, as it does, a few scribbles and some strong expressions crossed out and disregarded. From its concept and layout folio 138 is clearly not in letter form – it has no addressee, no sender's name and it is undated[10]. Folio 137 is a more polished version of the document, on the reverse of which is the name of one "Maitre Guillaume Fulconis". It has been suggested that Maitre Fulconis was a scribe who would have had the responsibility of transforming the draft into a formal ecclesiastical document which would have been resubmitted to Bishop d'Arcis for dating and signature[11].

The fact that this draft still exists and was not destroyed would indicate that the manuscript was never put into letter form, let alone sent to the Pope. It is also significant that neither document is dated, although there

The Later Years in Lirey

are a number of dating indicators in the documents[12]. The chronology of events is important.

On 28 July 1389 the Pope wrote to Geoffrey II de Charney giving his permission for the cloth to be publicly displayed and stating that Bishop d'Arcis should remain perpetually silent on the matter. The Memorandum refers to this letter from the Pope but states that Bishop d'Arcis had not yet seen a copy of the letter.

The Memorandum also refers to the successful legal appeal of the Bishop's dispute with Geoffrey II de Charnay to King Charles VI, a proceeding completed shortly before it was cited in the King's own letter to the *bailli* of Troyes dated 4 August 1389.

However, the Memorandum makes no reference to a report dated 15 August 1389 by the *bailli* of Troyes to the King that he had failed to repossess the Shroud from the Lirey church. Neither does it make any mention of a report by the King's First Sergeant dated 5 September 1389 that he had formally declared the Shroud to be royal property. These two events would undoubtedly have been mentioned by Bishop d'Arcis if his Memorandum had been drafted subsequent to them.

The document therefore appears to have been drafted in early August 1389. Yet a transcription of the Memorandum prepared at the start of the twentieth century gives a date of "end of 1389"[13].

This transcription of the d'Arcis Memorandum was prepared in 1900 by a well-known French historian and bibliographer of the time, Canon Ulysse Chevalier[14]. Canon Chevalier was a leader of a progressive faction of the Catholic Church that was seeking to advance an agenda of new ideas for the Church in the twentieth century and to do away with what was seen as "superstition". As such he was an unwavering opponent of any recognition of the Shroud's authenticity – a subject under discussion at the time following the photographs taken by Secundo Pio.

Canon Chevalier's transcription basically consisted of the heading of Folio 137 followed by the text of the first draft contained in folio 138, with the date "end of 1389" added. In fact the document as published by Canon Chevalier does not exist – it is a hybrid fabricated by Chevalier to establish a convenient date for the Memorandum[15]. In order to fully utilize the Memorandum to discredit the authenticity of the Shroud, Chevalier had to be able to demonstrate that the Memorandum had been received by the Pope, and responded to.

The Pope had written to Bishop d'Arcis on 6 January 1390, reiterating exactly what he had declared to Geoffrey II de Charny in his earlier letter of 28 July 1389 – namely that based upon facts that he knew to be certain, he would permit continued exhibitions of the Shroud in a prescribed manner, and that he would excommunicate the Bishop if he continued his opposition to this[16]. It was Chevalier's intention by using a date of "end of 1389" to show that this letter of the Pope had been written as a direct response to the Bishop's Memorandum.

In 1903 an English translation of Chevalier's transcription was published by an English Jesuit scholar, Fr Herbert Thurston. Fr Thurston took Canon Chevalier's distortions of the Memorandum a stage further by some selective editing. In particular he deleted one paragraph which seemed to contradict the "end of 1389" date. His objective seems to have been to reinforce Chevalier's contention that the Memorandum was only prepared at the end of 1389, that it was finalized in form and submitted to the Pope in Avignon[17]. This truncated translation was reproduced in full, undoubtedly without any knowledge of the editing work done by Canon Chevalier and Fr Thurston, by Ian Wilson in his book *The Turin Shroud*[17].

Although the Memorandum was never sent to the Pope, it is clearly an indication of Bishop d'Arcis's vehement opposition to the exhibition of the Shroud in Lirey. It seems that d'Arcis had also expressed his outrage to the King, for in August 1389 the King had ordered the *bailli* of Troyes to seize the Shroud so that it could be moved to another church in Troyes "under honest custody". The Dean of Lirey went so far as to deny the *bailli* access to the Shroud, although this could in fact have been construed as treason – refusing to obey an order of the King.

The key to Bishop d'Arcis's behaviour probably lies in the words of the King's command to the *bailli* of Troyes to "get the cloth and bring it to me, so that I might relocate it in another church in Troyes and place it under honest custody." Whose honest custody would the Shroud be under in Troyes, except that of the Bishop? Bishop d'Arcis wanted the Shroud for himself. There would seem to have been very good reason for this. In 1389 the nave of the unfinished Cathedral in Troyes collapsed. Records show that 1389 was the only year during that time that expenses in the Diocese were greater than income and the collapse of the nave was so significant that no major effort could be made for sixty years to complete the Cathedral[19]. Bishop d'Arcis therefore had a very strong financial motive to get hold of a relic that would bring pilgrims and income to his own Cathedral rather than to the small church in nearby Lirey.

The allegations made in Bishop d'Arcis's Memorandum are in fact contradicted by the evidence of the time and would indeed appear to be the outrage of a man who had his own purposes to serve rather than a reasoned statement of the facts. For example he quoted his predecessor, Henri of Poitiers, as having found the Shroud to be a forgery and having instituted proceedings against the Dean of Lirey and his accomplices. Yet there survives a document of Bishop Henri, dated 28 May 1356, only a few months before the death of the elder Geoffrey de Charny at the Battle of Poitiers, warmly approving the foundation of the church at Lirey and explicitly "praising", "ratifying" and "approving" the venture[19]. Bishop Henri also praised the "sentiments of devotion" of the elder de Charny. The Pope seemed to recognize Bishop d'Arcis's motives; he had already taken steps, as we have seen, to instruct d'Arcis to keep silent on the subject of the Shroud under pain of excommunication and to allow the expositions at Lirey to continue. D'Arcis himself produced no proof that the Shroud was a painting nor did he mention the name of the supposed forger. Bishop Henri's immediate successor at Troyes, Bishop Louis Raguier, had also maintained that the Shroud was genuine[21]. D'Arcis himself in the Memorandum stated that "it was on the previous occasion (of its exposition) declared to be the true shroud of Christ[22]", a clear indication that the elder de Charnay himself had believed this to be the case.

The expositions continued at least into 1390, as in June of that year a Papal Bull granted new indulgences to those who visited the Church of St Mary of Lirey and its relics[23].

De Charny's response to d'Arcis's anger had been to continue to take part himself in the expositions of the Shroud. D'Arcis wrote as follows in his Memorandum:

"Moreover, the knight (Geoffrey II de Charny), maintaining and defending this behaviour, by holding the said cloth with his own hands on a certain solemn feast, and showing it to the people with the observances above described.... But though I have earnestly and humbly cited the said knight and besought him that he would for a time suspend the exhibition of the said cloth until your Holiness could be consulted and should pronounce upon the matter, he paid no attention....[24]"

De Charny and the Dean of Lirey were clearly sufficiently convinced of the genuineness of the Shroud and of their support from the Pope that they could not only ignore the demands of the local Bishop, but even the orders of the King. In fact they probably had a very good understanding of the Bishop's motives, which would explain why they applied directly to the King and to the Pope regarding exhibiting the Shroud in Lirey, bypassing the Bishop.

D'Arcis's allegations that the Shroud was a fraud, and that this had been confirmed by his predecessor, Bishop Henri, can thus be seen as an attempt to discredit the owners of the Shroud rather than a reasoned statement of fact. One can surmise that had d'Arcis obtained possession of the Shroud he would quickly have found that it was not a forgery after all, but rather the genuine burial cloth of Christ, which would have brought direct and substantial financial benefit to his diocese.

The truth behind the d'Arcis Memorandum, therefore, would seem to be that it was an expression of the Bishop's anger at the expositions in Lirey, driven by his own financial needs. However, for reasons best known to himself – the Pope's express instruction to him to be silent would be one likely reason, but it may also be possible that he realised his allegations would not stand up to public scrutiny – he never had it written in a final letter form and it was never sent to the Pope.

Although the d'Arcis Memorandum can now be discredited as an accurate commentary on the Shroud and its history in Lirey, it is frequently used as the basis for claims that the Shroud has been known to be a forgery since its earliest appearance. The radiocarbon dating of the Shroud carried out in 1988 has reinforced this. Three pieces of cloth were cut from the Shroud and subjected to radiocarbon analysis in three separate and highly reputable laboratories. All three laboratories gave a date of about 1325, plus

The Later Years in Lirey

or minus 60 years, for the material which was tested. This dating has been the subject of much debate, which will be described in detail in Chapter 16. However, the coincidence of the radiocarbon dating and the dates of the events described in the d'Arcis Memorandum have provided strong grounds for belief that the Shroud might be a mediaeval forgery.

Geoffrey II de Charny died in 1398, leaving his lands in Lirey to his daughter Marguerite. The Shroud remained in the hands of the church in Lirey until 1418. There is no reason to believe that its exposition in the church there was interrupted at any time during the years 1389 to 1418. Bishop d'Arcis had been silenced by the Pope. The situation in France remained relatively stable for the first twenty years of this period and there was nothing like the anarchy of the 1360s during this time.

* * * * * * * * *

King Charles V of France had died in 1380 and had been succeeded by his young son Charles VI. There had been campaigns in Flanders during the early years of his reign and France had also become involved in other European affairs, particularly in Italy. In 1384 the King's uncle, Philip the Bold, Duke of Burgundy, had succeeded to the Counties of Artois and Flanders through his marriage to Margaret, the daughter and heiress of the previous Count. He also became Count of the Imperial County of Burgundy, uniting the County and Duchy for the first time in generations. He thus became an extremely powerful man, able to wield extensive influence in the affairs of France[25].

All was relatively well, however, until 1392 when King Charles VI had what appeared to be a nervous breakdown, followed shortly by a fit of madness. He was to suffer fits of madness for the remainder of his life[26]. This led to the passing of power into the hands of his uncle the Duke of Burgundy and his brother Louis, Duke of Orleans. Dissension between these two men and their heirs and partisans was to bedevil France for decades and to result in the renewal of the Hundred Years' War between France and England. Burgundy had the upper hand when the King was sick, but when he was lucid he relied heavily on Orleans. Philip the Bold died in 1404, to be succeeded as Duke of Burgundy by his son John the Fearless, who continued his father's hostility to the Duke of Orleans.

In 1407 the Duke of Orleans was murdered in Paris. The Duke of Burgundy confessed to the murder and fled from Paris, igniting a civil war between his supporters and those of the murdered Duke of Orleans. These latter were referred to as Armagnacs after the Count of Armagnac, the father-in-law of Charles of Orleans, the son and heir of Duke Louis. To some extent the conflict was between the north of France, represented by the Burgundians, and the south of France, represented by the Armagnacs.

A further step towards disaster occurred in 1411, when both sides started to negotiate with the King of England, Henry IV. Eventually the Armagnacs negotiated a treaty with Henry IV by which they would aid him in conquering the whole of the Duchy of Aquitaine. The Burgundian army at this stage became the army of France. Although a peace was concluded in 1412 the general situation in France had become one of crippled administration, high taxes and social unrest[27]. The King's son, the Dauphin John, had begun to take a leading part in the government of France and he sought to escape from the control of the Burgundians. Following pro-Burgundian riots in Paris the Armagnacs seized the city in 1414.

In England Henry IV had been succeeded by his son Henry V in 1413. Henry V, as Prince of Wales, had favoured an alliance with the Burgundians rather than his father's treaty with the Armagnacs. In 1415 he reasserted his claim to the French throne, based on the claim of Edward III that had set off the Hundred Years' War in 1337. The Duke of Burgundy promised to support Henry V against Charles VI and not to oppose his claim to the throne of France. Later that year Henry V landed in France, took the port of Harfleur and set off to march across northern France. His army met the French army at Agincourt and inflicted another decisive defeat on the scale of Crecy and Poitiers. At this stage Henry V had the tacit support at least of the Burgundians and the defeated French army consisted mostly of Armagnacs.

Warfare continued during 1417 and 1418, mostly in Normandy. In 1418 the Burgundians captured Paris and the King fell into their hands. The Dauphin Charles (his elder brother John had died in 1417) escaped and joined the Armagnac forces. By 1419 the situation was totally favourable to Henry V. The Court had fled into the Champagne region and the

The Later Years in Lirey

whole of Normandy was in Henry's hands. Adherents of the Dauphin did manage to murder Duke John of Burgundy in that year.

In May 1420 Henry marched to Troyes at the head of a strong force and signed a treaty of peace there by which he was recognized as Charles VI's heir to the throne of France. In June 1420 he married Charles VI's daughter, Catherine, at Troyes.

* * * * * * * * *

Much of this unrest in the kingdom and rivalry between princes must have passed Lirey by. Expositions of the Shroud could well have continued there in relative peace. Open warfare was confined to south-west France and then north-west and northern France. Champagne would have remained peaceful until the years after Agincourt. By the time of the Treaty of Troyes in 1420 it was once again right in the centre of events.

Marguerite de Charny's first husband, Jean de Baufremont, had been killed at the Battle of Agincourt. It can therefore be assumed that he was a supporter of the Armagnacs. The pre-eminence of the Burgundian allies of Henry V in the Champagne region may therefore have posed some threat to her personal security. In 1418 she remarried, this time to Humbert of Villersexel, Count de la Roche. Within one month of the wedding the Lirey canons handed over the Shroud to Humbert for safe-keeping[28], ostensibly because of danger from marauding bands. This can readily be believed in view of the spread of warfare from Normandy into neighbouring regions. The situation of insecurity in which Lirey found itself in the late 1350s was being replicated with the similar result that the Shroud was removed from the public eye and placed in safety. With this came the end of the Shroud's time at Lirey.

Chapter 6 –
From the de Charny Family to the House of Savoy

It was to be nearly a hundred years before the Shroud again found a permanent home. For much of the remainder of Marguerite de Charny's life it would travel with her wherever she went.

Marguerite's first husband would seem to have been an adherent of the Armagnac faction. Her second husband, Humbert of Villersexel, would appear to have been a loyal Burgundian. In any event Marguerite spent the remainder of her life in Burgundian territories and would seem at one stage to have enjoyed the personal protection of the Duke of Burgundy. Any fear she might have felt of a threat from the Burgundians because of her first husband's loyalties must have dissipated very quickly.

The document given by Marguerite de Charny and her husband to the canons of Lirey, when they assumed custody of the Shroud, named their castle of Montfort as the place where it would be kept. This document was dated 6 July 1418. This castle has generally been identified with a castle of Montfort near Montbard, in Burgundy south of Lirey. This castle had been built by Marguerite's father, Geoffrey II de Charny. However an investigation at this castle by the historian and former editor of the publication Shroud Spectrum International, Dorothy Crispino, found no evidence or tradition of the Shroud ever having been there. In fact, according to Sanna Solaro, writing in 1901, the Shroud was taken "to the castle of Montfort, also called Roche-Saint-Hippolyte". This was a castle in the personal domain of Humbert de Villersexel, at St Hippolyte sur Doubs, even deeper in the territory of Burgundy[1]. The Shroud is known to have been at St Hippolyte sur Doubs some time after 1418. Clearly there has been some historical confusion between castles of Montfort. According to

From the de Charny Family to the House of Savoy

seventeenth century records annual expositions of the Shroud were held in a meadow called Pre du Seigneurs on the banks of the River Doubs[2].

In Burgundy the Shroud was secure from the continuing warfare that bedeviled France for another thirty years after the Treaty of Troyes. Henry V of England died in 1422 – a major blow to the English/Burgundian cause in France. In any event the relationship between Henry and the Duke of Burgundy, Philip the Good, had deteriorated.

Charles VI of France died within months of Henry, and Henry's infant heir, Henry VI, was crowned King of France in Paris with his uncle the Duke of Bedford ruling in France in his name. This led to a situation where the Duke of Burgundy became virtually an independent sovereign within his own fiefs. Indeed, it became a political objective of the Dukes of Burgundy to have their territories recognized as a kingdom independent of both France and the Holy Roman Empire in Germany. As a result of this France was effectively divided into three parts – the fiefs of the Duke of Burgundy, the domain of Henry VI and the regions acknowledging the Dauphin, now Charles VII, as King of France.

The Dauphinists remained active in the Champagne region with the result that unrest continued there for many years. A great deal of brigandage occurred – some was the result of partly disciplined bands and some from the activity of opportunists out to make what they could from living off the land.

Charles VII himself was an inactive and incompetent king who surrounded himself with venal advisors. The Duke of Bedford consolidated Henry's claims in France with a victory over the Dauphinist forces at the Battle of Verneuil in 1424[3].

The situation remained a virtual stalemate for ten years. In 1428 Orleans, a Dauphinist stronghold, was besieged by an English army. A French force led by the visionary Joan of Arc entered Orleans and raised the siege. Joan firmly believed that Charles VII should be crowned at Rheims, the traditional coronation place for the Kings of France, and that the English should be expelled from France. She proposed a march from Orleans to Rheims that would involve an expedition across Champagne. This was carried out in 1429 and the coronation of Charles VII duly took place. In 1430 Joan was captured by the English and she was burned at the stake as a heretic in Rouen in 1431[4].

In the meantime both the English and the Dauphinists had been carrying out negotiations with the Duke of Burgundy. Burgundy decided to continue his alliance with the English, for the time being at least. However he eventually signed a truce with Charles VII in 1434. In 1435 he abandoned his English alliance and signed a Treaty with Charles. This effectively sounded the death knell for English rule in France and heralded the approaching end of the Hundred Years' War, although the English did manage to continue the struggle for another eighteen years.

The French countryside was again ravaged by marauders over the next ten years and Champagne was affected. The ravages of the *Ecorcheurs*, as they were called, matched in ferocity and brutality the ravages of the mercenary bands of the 1350s and 1360s. While France suffered, the territories of the Duke of Burgundy once again escaped relatively lightly.

Royal authority was gradually reasserted in France. A truce between the French and the English was signed in 1445 and lasted for four years. When hostilities resumed Normandy was eventually recaptured by the French. Guienne was then conquered and Bordeaux was finally captured in 1453, marking the end of the Hundred Years' War[5].

Humphrey of Villersexel died in 1438, possibly killed in battle[6]. By 1443 the situation in Champagne would seem to have returned to something approaching normality as the canons of the church in Lirey petitioned Marguerite to return the Shroud to them, as it belonged to them. Marguerite refused and in a deposition stated that the Shroud was *"conquis par feu messier Geoffroy de Charny"*[7]. This is a curious phrase which has caused a degree of puzzlement to historians. An explanation has been suggested by Patrick Farrell of Rossshire in Scotland. His suggestion is that *feu* refers to Geoffrey de Charny and is the same as in the well-known phrase *feu le roi, vive le roi*, meaning "the king is dead, long live the king". The derivative of *feu* in this sense is not "fire" but the Latin word *fatutus* meaning deceased. In this context *conquis* means no more than "obtained"[8]. The meaning of Marguerite's phrase would therefore seem simply to be that she obtained the Shroud through the death of Geoffrey de Charny; in other words it was hers by right of inheritance.

From this time on Marguerite seems to have traveled, taking the Shroud with her. All her travels appear to have been within the territories of the Duke of Burgundy, which, at the time, included Flanders and Artois in north-eastern France and modern-day Belgium. In 1448 Marguerite is recorded in the Archives of Mons as having in her care "what is called the Holy Shroud of Our Lord" and entering Mons where she ordered French wine[9]. The following year she is recorded as exhibiting the Shroud at Chimay in the diocese of Liege. By 1452 she had returned to Burgundy where she showed the Shroud at the castle at Germolles, near Macon[10]. The castle at Germolles was one of the personal residences of the Duke of Burgundy. The fact that Marguerite appears to have been staying there would seem at first sight to indicate that, at the time, she enjoyed the patronage of the Duke. Yet she decided the following year to cede ownership of the Shroud to the House of Savoy rather than to the House of Burgundy[11].

* * * * * * * * *

At the end of the Second World War the House of Savoy was the oldest royal house of Europe still on a throne. The founder of the family was a Burgundian noble called Humbert, known as "the Whitehanded", who lived during the first half of the eleventh century. He may have been the great-grandson of the Holy Roman Emperor Otto II, but this is unproven. In 1003 he acquired, possibly from the Emperor Conrad II, certain Alpine territories as a feudal lord, as a result of which he became known as the Count of Savoy[12]. Humbert consolidated his influence in Savoy by supporting Conrad II in his claim to the Kingdom of Upper Burgundy and was rewarded with large gifts of land[13]. The family increased its landholdings and influence in the next generation, when Humbert's son, Odo (or Otto), married the heiress of the Count of Turin and gained for his descendants through this marriage possession of a large part of Piedmont in Italy, including Turin itself.

The Counts of Savoy maintained their close relationship with the Holy Roman Emperors and continued to prosper as a result. Count Thomas I was imperial vicar to the Emperor Frederick II and further enlarged his family's possessions.

Count Amadeus V was known as "the Great"[14] and became a prince of the Empire. His successors continued to expand their territories and his son Amadeus VI made a law that his territories should never be divided. The daughter of Amadeus V, Agnes, married the Count of Geneva and had a number of children. One of her sons became Count Amadeus III of Geneva; his cousin Aymon of Geneva married, as her second husband, Jeanne de Vergy, the widow of Geoffrey de Charny. There were numerous family ties through marriage between the House of Savoy and the House of Geneva, and the second marriage of Jeanne de Vergy therefore established another link between the de Charny family and the House of Savoy[15]. There was already a link in that Marguerite de Charny was a direct descendent of an early Count of Savoy, Humbert II, who lived in the eleventh century; she was therefore herself a distant cousin of the Savoys[16].

By the time of Marguerite de Charny the House of Savoy held wide territories on both sides of the Alps, including Geneva, Nice, Turin and Piedmont. The capital of their territories was at Chambery in France. In 1416 the Count of Savoy had been raised to the rank of Duke by the Emperor Sigismund, further enhancing the power and influence of the Savoy rulers.

* * * * * * * * *

The events surrounding the transfer of ownership of the Shroud from Marguerite de Charny to the House of Savoy involve a web of politics and intrigue at the Savoy Court.

Duke Louis I of Savoy had married Anne de Lusignan, the daughter of Janus, the nominal King of Jerusalem, Cyprus and Armenia. The de Lusignans were the direct descendants and heirs of the de Lusignans who had been Kings of Jerusalem at the time of the Third Crusade in the late twelfth century. Anne was praised as the most beautiful and most gracious lady in the world; she was also extremely extravagant and self-willed. Pope Pius II described her as "a woman incapable of obeying married to a man incapable of commanding."[17]

The heir of Marguerite de Charny's second husband, Humbert de Villersexel, was his nephew Francois de la Palud, a Savoy nobleman who had also inherited lands in Burgundy and who was a protégé of the Duke of Burgundy. He was a notable soldier who had lost his nose in battle. As a result of this he wore a false nose made of silver and was known as the Knight with the Silver Nose.

Francois had plotted to murder another Savoy noble, Jean de Compey, who happened to be a favourite of the Duchess Anne. Despite her annoyance, her husband and his father, the former Duke Amadeus VIII (who had abdicated in 1440), declined to take any action against Francois. They had sound political reasons at the time for this prudence. Duke Amadeus died in 1451 and Louis, no doubt under the influence of his wife, then banished Francois from Savoy and confiscated his property. He also ordered the destruction of Francois's castle of Varambon.

Francois fled to his lands in Burgundy and appealed for support to both Duke Philip the Good and King Charles VII of France. The King already had reason to dislike Duke Louis, as his son, also named Louis (later Louis XI of France), had married the ten-year old daughter of Louis of Savoy, Charlotte, without informing his father[18]. King Charles and the Dauphin Louis were at loggerheads and this marriage had given Charles cause for resentment against Louis of Savoy. Charles made threatening military gestures against Louis, the outcome of which was that Louis had to agree to withdraw charges against Francois de la Palud and return his property to him, or to compensate him for what had been destroyed (the castle of Varambon).

At this time Marguerite de Charny was at Macon in Burgundy. For some reason she decided to put in a claim for Francois's property, including compensation for the destroyed castle of Varambon. Her claim possibly arose from her marriage to Francois's uncle. Duke Louis was in a quandary. He had promised the King of France that he would return Francois's property to him, but his wife, Anne, was still adamant that Francois had to be punished. Marguerite's claim gave him a way out. On 22 March 1453 Duke Louis reached an agreement in Geneva with Marguerite giving her the rights to the Varambon estate. At the same time she ceded the Varambon estate back to the Duke in return for the estate of Miribel near Lyons. This agreement is frequently assumed to be that by which Marguerite handed over the Shroud to the House of Savoy, but there is in fact no reference to the Shroud in it and it does seem to be more related to the problems concerning Francois de la Palud. On the other hand it has to be noted that the day after this agreement, 23 March 1453, Duke Louis issued a medal in honour of the Shroud[19]. Whether this was in recognition of the actual transfer of ownership of the Shroud or was merely a recognition of Marguerite's association with the Shroud must remain a matter of conjecture.

If Duke Louis thought that this would solve his problem with King Charles VII, he was wrong. The King continued to demand that Louis abide by his agreement to pardon Francois de la Palud and to compensate him, and eventually the Duke had to pay. At the same time (this was in 1455), for some reason that is not clear, Marguerite renounced Miribel and accepted the estate of Flumet. To complicate things even further, the Duke had forgotten that he had previously promised Flumet to one of his counselors. The sons of the latter fought a claim to Flumet and won it, leaving Marguerite the loser[20].

The question arises, just where did the Shroud fit into all of this?

The role of the Duchess Anne cannot be overlooked. She was a descendant of the Crusader Kings of Cyprus and had been brought up there. At that time the Orthodox Church in Cyprus still celebrated 16 August as the feast of the coming of the Cloth of Edessa to Constantinople, despite its disappearance in 1204. Although a Catholic, Anne could not have been unaware of these celebrations. She may well have recognized an affinity between the Cloth of Edessa and the Shroud of Lirey. She may well have believed that the two were indeed one and the same[21]. She may also have felt that she had some claim to ownership of the Shroud through her descent from the Crusader Kings of Jerusalem. Her extreme piety, combined with this background, may well have given her the desire to possess the Shroud. As has already been mentioned, she was a headstrong and self-willed person.

This is obviously very speculative, but it does seem quite likely that Anne played a key role in obtaining the Shroud for the House of Savoy from Marguerite de Charny.

Marguerite herself had no children or direct heirs. Her closest relatives were her half-brother Charles de Noyers, the son of her mother's second marriage, her cousin and godson Antoine Guerry des Essars, to whom she bequeathed her lands at Lirey, and her late husband's disreputable nephew Francois de la Palud. She clearly felt that none of these were suitable owners for the relic she had inherited from her father and grandfather. On the other hand she had both a financial and a family relationship with the House of Savoy.

Somewhere in all this – the dealings over the property of Francois de la Palud, the ambitions of the Duchess Anne and the lack of any other suitable heirs for the Shroud – Marguerite transferred ownership of the Shroud to the House of Savoy. It may not have been in 1453 and it may not have had anything to do with the transfer of the Miribel estate to her.

Marguerite de Charny's decision to transfer ownership of the Shroud to the House of Savoy rather than, say, to the House of Burgundy, was a critical factor in the long-term survival of the Shroud. Although she could not foresee the future, the guardians she chose continued to increase in prosperity and power until they became the Kings of Italy. Five hundred years after taking ownership of the Shroud the Savoys were still one of the leading royal houses of Europe. The House of Burgundy had perished before the end of the fifteenth century.

Despite the change of ownership of the Shroud the canons of Lirey continued to seek its return to what they clearly considered to be its rightful home. In 1457 Marguerite was threatened with excommunication by the church in Lirey if she did not return the Shroud and then excommunicated when she failed to return it. Two years later her half-brother, Charles de Noyer, reached an agreement with the canons of Lirey whereby they would accept compensation for the loss of the Shroud and lift the excommunication on Marguerite. In 1464 Duke Louis of Savoy agreed to pay the canons an annual rent from the revenues of one of his castles, near Geneva. This agreement was drawn up in Paris and is the first surviving document to show that the Shroud had become the property of the Savoys. It also clearly shows the Lirey Shroud and that subsequently owned by the Savoys to be identical. It specifically noted that the Shroud had been given to the church of Lirey by Geoffrey de Charny and that it had been transferred to Duke Louis by Marguerite de Charny[22]. Meanwhile, Marguerite had died in 1460.

The Savoys appear to have been reluctant to meet their commitments to the church of Lirey. Nine years after the agreement was drawn up the rent was eight years in arrears and the canons of Lirey again pressed for the Shroud's return if payment was not forthcoming. In 1482 they made what appears to be one last effort to obtain some form of compensation and then they seem to have given up what had clearly become a lost cause[23].

Chapter 7 – To the Present Day

On its becoming the property of the House of Savoy the role of the Shroud subtly changed. Under the ownership of the de Charny family, the Shroud had been seen as no more than the image of Christ on a cloth. At times it was represented as the true burial cloth of Jesus; at other times the image was said to be merely a likeness of the crucified Christ. It was seen as a relic that had attracted pilgrims over the years – from its first exposition in Lirey by Geoffrey I de Charny in about 1355 to the exposition by Marguerite de Charny at the Castle of Gemolles in Burgundy nearly one hundred years later, in 1452.

For the House of Savoy, however, it rapidly became the family palladium – a divine protective device to be invoked at times of difficulty or danger. In its first decades with the Savoys the Shroud had no fixed home. It was carried about by the family on all their travels, much as it had been by Marguerite de Charny in her later years. The Savoys, however, appeared to use it to safeguard them against the dangers of their journeys[1].

The Shroud also grew in reputation in the ownership of the Savoys. It was no longer just the property of an obscure member of a relatively minor noble family. It was now owned by one of the leading families of Europe. Louis and Anne were noted for their piety. Louis himself was deeply religious. He surrounded himself with Franciscan friars and had a Franciscan as his confessor. When he died he was buried wearing a Franciscan habit[2]. Anne's own piety has already been noted.

In 1464 the Franciscan theologian Francesco della Rovere (later Pope Sixtus IV) referred to the Shroud in the following terms in a work entitled *The Blood of Christ*:

"... the Shroud in which the body of Christ was wrapped when he was taken down from the cross. This is now preserved with great devotion by the Dukes of Savoy, and it is coloured with the blood of Christ[3]."

The piety and devotion to the Shroud of Louis and Anne were inherited by their successors. Louis was succeeded as Duke of Savoy by his son Amadeus IX. Amadeus was later said to have instituted the cult of the Shroud in the Sainte Chapelle at Chambery, at that time the capital of the Savoy territories. His wife, the Duchess Yolande, founded a convent of Poor Clare nuns at Chambery[4].

During the 1470s and 80s the Shroud traveled extensively with the Savoys. Duke Amadeus IX lived only a short time, but his wife Yolande continued to act as Regent for their young son Duke Philibert I. During this time the Shroud was moved across the Alps, firstly to Vercelli, approximately half way between Turin and Milan, and then a few years later to Turin itself. Shortly after that it was back in Chambery, and it then crossed the Alps yet again for exhibitions in the Turin area. By 1485 Philibert's brother Charles had become Duke. He married Blanche de Montferrat, who became known as the Duchess Bianca and who followed in the footsteps of previous Duchesses in her piety and veneration of the Shroud. The Shroud was carried about by Charles and Bianca on their Court journeys from castle to castle. That year a clerk, Jean Renguis, was recorded as having been paid two ecus "in recompense for two journeys which he made from Turin to Savigliano carrying the Shroud[5]."

On Easter Sunday 1488 the Shroud was exhibited at Savigliano and on Good Friday six years later, in 1494, it was exhibited once again at Vercelli. This was recorded by the secretary to the Duke of Mantua, one Rupis, who reported to the Duke as follows:

"A sudarium was exhibited, that is, a sheet in which the body of Our Lord was wrapped before being laid in the tomb, and on which his image can be seen outlined with blood – both front and back – and it looks as though blood is still issuing[6]."

A theory has been proposed by Lynn Picknett and Clive Prince in their book *Turin Shroud – in whose Image?* that the Shroud that exists today was created in 1492 by Leonardo da Vinci through a photographic process. It is their hypothesis that the Shroud of Lirey that the Savoys obtained from Marguerite de Charny was nothing but a painting and that it became necessary for a new and more realistic looking Shroud to be created. In support of their claim they suggest that there is no record of the Shroud having been displayed or even seen in the 40 years prior to the 1494 exposition at Vercelli[7]. Clearly this is incorrect, as can be seen from the record of travels and expositions of the Shroud during the 1470s and 1480s. It is not unreasonable to suggest that there must have been many people who attended both the 1488 Savigliano exposition and that in Vercelli six years later. There would certainly have been a number of members of the Savoy Court at both events. If there had been a switch in the cloth being exhibited surely someone would have noticed it!

Picknett and Prince's contention that the faking of the Shroud by da Vinci was done at the instigation of Pope Innocent VIII also makes little sense. They suggest that it was a cynical publicity exercise on the part of the Pope[8], who wanted another, better Shroud with which to pull in the crowds[9]. However, they themselves acknowledge that Innocent VIII was one of the weakest and most ineffectual of all the fifteenth-century popes[10]. He was in Rome while the Shroud was in Savoy. There is no evidence of Pope Innocent VIII ever taking any interest whatsoever in the Shroud; it was never even remotely under his control and if he had wanted to carry out any publicity exercises – particularly ones that would have attracted pilgrims and hence income – he no doubt had far better opportunities available to him in Rome[11].

Any such conspiracy to create a new Shroud would also have had to involve as one of the principal conspirators the Duchess Bianca of Savoy. Again there is no evidence of this religious woman ever having been involved in any conspiracy or plot of this nature and it is certainly not clear what her motive would have been in promoting the creation of a replacement Shroud.

It is worth noting, however, that two Popes of the time did take a particular interest in the Shroud. Pope Sixtus II, as the theologian Francesco della Rovera, specifically wrote about the Shroud being the genuine shroud of Christ. This was before the supposed switch of the Shroud. Pope Julius II, who happened to be the nephew of Sixtus II, went so far as to assign the Shroud its own feast day. This was in 1506, after the supposed switch[12]. If

Innocent VIII had been a party to the faking of the Shroud by Leonardo da Vinci, it is certain that his successors, including Julius II, would have known of it. (There were two popes between Innocent VIII and Julius II.) For Julius II to have given the Shroud its own feast day, knowing that it was a fake produced by da Vinci, would have been an act breath-taking in its cynicism even by the standards of the time.

Finally, at the time that da Vinci is said to have worked on creating the new Shroud he was in fact deeply involved in a project for an equestrian statue of Duke Francesco de Medici of Florence. This project had been commissioned in 1489. Between 1491 and 1493 da Vinci was kept busy considering the problem of casting the horse and certainly spent most of his time in Florence[13]. There is no reason why he should have interrupted his work on this project to become involved in the affairs of Savoy.

*   *   *   *   *   *   *   *   *

The Shroud finally found a permanent home in 1502, nearly a hundred years after it had been removed from Lirey. At this time Philibert II was Duke of Savoy and his Duchess was Margaret of Austria. The Duchess Margaret was as devoted to the Shroud as previous duchesses had been. It was her decision to give it a permanent home in the ducal church in Chambery. In June that year the Shroud was solemnly taken to the Royal Chapel where it was displayed on the High Altar before being deposited in a special cavity behind the altar[14]. Four years later Pope Julius II, as well as instituting an official feast day for the Shroud, gave approval for this chapel to be called the Sainte Chapelle of the Holy Shroud, giving it the same status as the Sainte Chapelle in Paris, which had been built in the thirteenth century by King Louis IX of France to house the Crown of Thorns[15]. The Shroud remained safely in this chapel for the next thirty years.

During this time at least one copy was made of the Shroud. Of the various copies that have been made, one of the best-known is the one kept in the Church of St Gommaire at Lierre in Belgium. This may have been made for the Dowager Duchess Margaret of Savoy, who had become Regent in the Netherlands and had set up a Court there. The Shroud copy at Lierre is notable for clearly showing the "poker holes" that may date back many centuries on the original Shroud[16].

To the Present Day

Regular public and private expositions of the Shroud were held during this period. One notable visitor was King Francis I of France, who walked from Lyons to Chambery to give thanks to God for success in battle. Public expositions took place on the castle walls so that the Shroud could be clearly seen by pilgrims.

The Duchess Margaret died in early 1531. In her will, which had been drawn up in 1508, she had stipulated that she wished to leave a portion of the Shroud to her church of Brou at Bourg-en-Bresse[17]. Her will was duly executed but there is no contemporary record of a portion of the Shroud having been removed as stipulated by her. However there are indications that this may have been done. Margaret's nephew and ward was the Holy Roman Emperor Charles V, a leading figure in the fight of the Catholic Church against the Protestant Reformation. He is certainly likely to have ensured that all of her bequests were carried out, including one of such religious significance. A Shroud replica commissioned forty years later for Don Juan of Austria, the son of Charles V, by Pope Pius V, clearly shows a corner of the Shroud as having been excised.

Recent researchers have suggested that a section of the cloth measuring five and a half inches by three and a half inches was removed in 1531 as a result of the Duchess Margaret's bequest. This was then replaced by the most skilled craftsmen of the time with what was essentially an invisible patch. This would be a matter of very minor historical interest if it were not for the fact that the patch in question is adjacent to where the samples were taken from the Shroud in 1988 for the radio-carbon dating tests. This has an important bearing on the validity of the radio-carbon dating which will be addressed in more detail in Chapter 16[18].

On 4 December 1532 a major fire broke out in the Sainte Chapelle in Chambery. The grille covering the cavity where the Shroud was kept had four locks, the keys to which were kept by separate officials. As a result of delays in finding these officials a blacksmith, Guillaume Pussod, had to prise open the grille to rescue the Shroud. Unfortunately by this time the silver casket in which the Shroud was kept had started to melt. When it was opened it was found that molten silver had caused the Shroud to partially catch fire. Although this was quickly put out by water, the Shroud had suffered serious damage[19].

The Shroud had been folded into forty-eight in its casket. When it was opened out into its full length, the effect was similar to that achieved by cutting a hole in a folded piece of paper. The cloth was extensively scarred with scorch-marks, burn holes and water-damage marks. Fortunately the central image on the Shroud was virtually untouched[20].

Two years elapsed before repairs were made to the fire damage. Then in 1534 the Shroud was taken to the Convent of the Poor Clares in Chambery, where the Mother Superior, Louise de Vargin, and three other nuns worked for two weeks to repair and patch the Shroud. Once the repairs had been completed, the Shroud was returned to the Savoys' castle[21].

The patches put on the Shroud by the Poor Clare nuns were removed during restoration work on the Shroud carried out in 2002[22].

* * * * * * * * *

The major European conflict of the fourteenth and fifteenth centuries had been the Hundred Years' War between England and France. The first half of the sixteenth century would be marked by the struggle between France and the Holy Roman Empire, particularly in Italy. Like so much mediaeval European warfare, this conflict had its roots in the dynastic and territorial ambitions of the leading monarchs of Europe. One of the factors involved was the ambition of the fifteenth-century Dukes of Burgundy to establish their territories as an independent kingdom, separate from France and the Empire. Centuries previously there had been such a "middle kingdom" in Europe, the Kingdom of Lotharingia, ruled by one of the sons of Charlemagne, Lothar, but it had been short-lived. It was the ambition of the Dukes of Burgundy and their successor, the Emperor Charles V, to revive this kingdom[23].

The line of the Dukes of Burgundy had come to an end in 1477 with the death of Duke Charles the Bold at the Battle of Nancy, against the French King Louis XI. This was the culmination of a war between France and Burgundy that had lasted some twenty years. At this time the Burgundian territories encompassed the Duchies of Burgundy, Nevers and Artois in France, and the Counties of Burgundy, Lorraine, Luxembourg, Hainault, Brabant and Flanders in the Empire. In the subsequent Peace of Arras signed in 1482, France took possession of the French Burgundian territories, while the Imperial Burgundian territories passed to Maximilian of Hapsburg, King of the Romans, who had married the daughter of Charles the Bold[24].

One of the key events of the early sixteenth-century was the rapid rise in the power of the Hapsburgs. They had held the Imperial throne for some time, but their power in European affairs had remained limited. In 1490 Maximilian was a refugee from Vienna, which was then in the hands of the Hungarians. By 1508 he had been confirmed as Emperor and his power grew to the extent that he even suggested that he should be elected Pope. After the death of his first wife, Mary of Burgundy, he had married Bianca Sforza, a member of the ruling family of Milan[25].

The son of Maximilian and Mary, Philip, had married Joanna the Mad, the daughter of Ferdinand and Isabella of Spain. They in turn had a son, Charles, who was not only the great-grandson of Duke Charles the Bold of Burgundy but also the heir to both the Hapsburgs and to Castile and Aragon in Spain. Charles was elected Emperor in 1519 and his empire included not only the German and Italian possessions of the Hapsburgs but also the European and overseas possessions of Spain. His empire stretched from the Philippines in the East to Peru in the West and encompassed an area three times as large as the Roman Empire at its height under the Emperor Trajan[26].

War in Italy broke out in 1494 when King Charles VIII of France sought to claim the inheritance of the former French Angevin rulers of Naples. (The Angevins had ruled Naples from 1268 – 1435, when they had lost the crown to Alfonso V of Aragon.) Charles invaded Italy, starting a period of warfare that lasted from 1494 until 1559. He also made a claim to the Duchy of Milan through his cousin Charles of Orleans, who was a descendant of the previous Milanese ruling family, the Viscontis.

With the conflicting claims in Italy of the Hapsburg Emperors and the Valois Kings of France, the scene was set for the rivalry between Francis I of France, who came to the throne in 1515, and the Emperor Charles V, who had been elected Emperor in 1519. To exacerbate the rivalry, Francis had himself sought the Imperial Crown, to be defeated by Charles. Charles, in his turn, sought to re-constitute the Burgundian territories of his great-grandfather and join them together in a realm running between the Rivers Rhine and Rhone and stretching from the Mediterranean and the Alps to the North Sea[27].

Charles V fought five wars with France. At first Savoy, despite its geographical location between the territories of the two rivals, was not affected, but eventually French troops invaded Savoy in 1535[28].

Milan was a key strategic point in this struggle. In addition to the conflicting dynastic connections to Milan, this Duchy was an indispensable link in an Empire stretching from Spain and Naples to the Baltic and the North Sea[29]. Francis saw it as a lever by which he might break up the Empire.

In 1525 Charles defeated Francis at Pavia and captured him. To obtain his release Francis had to promise to give up all his territorial claims, which he did by the Treaty of Madrid in 1526. Once released of course, in true Renaissance style, he immediately resumed his claims and renewed the war. Through the League of Cognac in 1526 he forged an alliance of Milan, Florence, Venice and the Papal States. In response to this Charles seized Milan and then captured and sacked Rome in 1527. Another peace was signed at Cambrai in 1529 by which France kept Burgundy but renounced its claims to Milan, Naples, Artois and Flanders.

Francis invaded Savoy in 1535 and Duke Charles III of Savoy and his family abandoned Chambery and retired to Piedmont, taking the Shroud with them. That year it was exhibited in Turin and the following year in Milan[30]. Further French invasions caused the Shroud to be taken to Vercelli for safe-keeping. In 1553 French troops sacked Vercelli, but the Shroud had been hidden away safely.

Savoy's fortunes were restored by the successor of Duke Charles III, Emmanuel-Philibert, who became known as "Iron-Head". He allied himself firmly with the Empire and defeated the French at the Battle of St Quentin. As a result of the Treaty of Cateau-Cambresis in 1559, Emmanuel-Philibert recovered his territories in Savoy and had Savoy's position guaranteed as a buffer state, while Spain gained hegemony over the rest of Italy[31]. This treaty ended the wars between France and the Empire. From that time France's orientation turned towards the Netherlands and the Rhineland, although she did retain fortresses in Savoy on the Italian slopes of the Alps.

* * * * * * * * *

The Shroud was moved permanently to Turin in 1578 by Duke Emmanuel-Philibert. Over the years that followed it was exhibited regularly, to large crowds of pilgrims. These expositions frequently corresponded with special occasions, such as marriages of members of the House of Savoy. Numerous copies of the Shroud were also made.

In 1694 a new black lining cloth was attached to the Shroud and some of the patches put on by the Poor Clare nuns to repair the fire damage of 1532 were repaired or replaced.

The fortunes of the Savoys also continued to rise. In 1713 Duke Victor Amadeus II obtained the crown of Sicily, which he exchanged eight years later for the crown of Sardinia, which he could rule more easily from Turin. During the Napoleonic Wars the Savoys retreated to Sardinia, leaving the Shroud in Turin. During this time it was shown privately to Pope Pius VII, who was en route from Rome to Paris to crown Napoleon as Emperor of the French. Pius again presided over an exposition of the Shroud in 1815, to mark his return to Italy after the defeat of Napoleon.

During the nineteenth century Sardinia absorbed its sister Italian states, including the Kingdom of the Two Sicilies. In 1861 Victor Emmanuel II was declared King of Italy.

In 1868 Princess Clotilde of Savoy replaced the black lining cloth with one of crimson taffeta sewn the length of one side of the Shroud.

In 1898 an eight-day exposition of the Shroud was held to mark Italy's fiftieth anniversary as a kingdom ruled by the House of Savoy. The photographer Secondo Pio took the photographs that started the scientific study of the Shroud. The twentieth-century history of the Shroud is primarily a history of these studies and the frequently controversial results that they have yielded.

Mark Oxley

During the Second World War, as had happened on so many previous occasions when warfare threatened, the Shroud was removed to safety. Following the outbreak of the war in September 1939 the Shroud was secretly taken to the Benedictine Abbey of Montevergine, in the province of Avellino, north-east of Naples. It remained there until October 1946, when it was exhibited to the monks, most of whom had not even been aware of its presence in the Abbey. It was then returned to Turin and its housing in the Royal Chapel. By this time the people of Italy had voted for a Republic and the Savoys were not only no longer Kings of Italy but they had no official status there at all. The last Savoy King, Umberto II, had to go into exile and he lived out most of the rest of his life in Portugal.

In 1983 ownership of the Shroud passed from the House of Savoy. Ex-King Umberto II of Italy died and bequeathed the Shroud to Pope John Paul II and his successors, with the proviso that it should remain in Turin[32].

In April 1997 fire once again threatened the Shroud. A cryptic warning was given to the local police that a fire might occur at Turin Cathedral. Shortly afterwards the Cathedral's fire alarms went off, but the Royal Chapel was already blazing. Had it not been for the bravery of fireman Mario Trematore the Shroud would have been destroyed. He saw the new display case to which the Shroud had recently been removed and succeeded in smashing the glass of the case. He was able to reach in and pull out the Shroud's silver casket, which was rapidly taken to safety. On this occasion, unlike in 1532, the Shroud suffered no damage from the fire[33]. Although the fire appears to have been a deliberate act of arson, nobody has ever been accused of the crime.

In 2002 a conservation programme lasting thirty-two days was carried out on the Shroud by the Swiss textile conservator Mechthild Flury-Lemburg and her assistant Irene Tomedi. They removed the backing cloth that had been put in place by Princess Clotilde in the nineteenth century, as well as the patches that had been used to repair the Shroud after the fire of 1532. Debris and other material behind each patch were gathered, recorded and preserved in special canisters for future scientific examination. A new backing cloth, selected for its not having been treated with dyes, starches, bleaches and other potential contaminants, was sewn on to the Shroud[34].

Today the Shroud is kept in a side chapel to the left of Turin Cathedral's high altar, beneath what looks like a rather plain, box-like altar, a continuing challenge to scientists, thinkers and laymen of the twenty-first century.

* * * * * * * * *

As a historical footnote, it is worth recording how many popes have at least venerated the Shroud if not actually recognizing it as the authentic burial cloth of Jesus Christ.

The first pope to be associated with the Shroud since its first appearance in Lirey was the Anti-Pope Clement VII, the pope so closely associated with the d'Arcis Memorandum. On 28 July 1389 Clement wrote a letter to Geoffrey II de Charny declaring that, based upon facts known to him *ex certa scientia*, the Shroud could be publicly displayed and that Bishop d'Arcis must remain perpetually silent on the matter. He repeated this in a letter to Bishop d'Arcis dated 6 January 1390[35].

It has frequently been pointed out that Clement VII was the nephew of Jeanne de Vergy's second husband, Aymon of Geneva, and hence closely connected to Geoffrey II de Charny. In seeking papal authority to exhibit the Shroud, Geoffrey by-passed Bishop d'Arcis and made a direct approach to the Papal Legate, Cardinal de Thury, who obtained the necessary permission from Clement without any difficulty. The tone of Clement's letters would seem to indicate that he had some knowledge of the real origins of the Shroud[36]. The question of where the elder Geoffrey de Charny did obtain the Shroud will be addressed in detail in the next two Chapters. However it does indeed seem possible that Clement had some knowledge of the Shroud's origins and based his decision on that. It can be speculated that Clement, before being elected Pope, visited his relative Aymon de Geneve and his wife Jeanne de Vergy and learned of the Shroud's history from her. However no record exists today, or certainly no known record, of any papal knowledge of the Shroud's origins.

Although permitting the exhibition of the Shroud, Clement did not officially recognize it as the true burial cloth of Christ. Indeed de Charny was instructed to make it clear that the cloth was not the true shroud of Christ[37]. In direct contradiction of the papal instructions de Charny went ahead to suggest that the Shroud was authentic and it would seem at least possible that he did this in the knowledge that, despite his written instructions, Clement shared this belief. Certainly in June 1390 he issued a Papal Bull granting new indulgences to those who visited the church in Lirey and its relics[38].

Several centuries later, in the mid-eighteenth century, Pope Benedict XIV referred to the Holy Shroud being preserved at Turin and specifically referred to Popes Paul II (1464 – 71), Sixtus IV (1471 – 84), Julius II (1503 – 13) and Clement VII (1523 – 34) as all having borne witness that the Shroud is the same as the one "in which our Lord was wrapped"[39]. (This Clement VII is not the fourteenth century anti-pope in Avignon but a later pope who is recognized as a legitimate pope.)

Pope Paul II had in 1467 elevated the status of the Chambery chapel where the Shroud was kept to a collegiate church, obviously in recognition of its status. As the Cardinal Francesco della Rovere, Pope Sixtus IV had written of the Shroud, "On this Shroud we see the image of Jesus Christ traced in his very blood." In 1506 Julius II had instituted a Feast of the Holy Shroud, complete with proper Mass and Office, to be celebrated on 4 May. Initially this Feast was just for Chambery, but in 1514 Pope Leo X extended it throughout Savoy[40].

In 1571 Pope Pius V commissioned two copies of the Shroud, one of which he gave to Don John of Austria, the son of the Emperor Charles V. In 1582 Pope Gregory XIII extended the Feast of the Holy Shroud throughout the entire dominions of the Duke of Savoy, with the further grant of a plenary indulgence to pilgrims[41]. Pope Pius VII knelt to venerate the cloth when he visited Turin in 1804, on his way to Paris, and he personally displayed it at an exposition ten years later.

In the twentieth century, Pope Pius XI told the Archbishop of Turin, Cardinal Fossati, in a private audience in 1931, "Be entirely at ease. We speak now as a scientist and not as Pope. We have made a personal study of the Holy Shroud and are convinced of its authenticity. Objections have been raised but they do not hold water.[41]" Two years later, at the same Pope's request, the Shroud was exhibited as part of the celebrations for the Holy Year. Pius XI had a particular interest in the Shroud, as has been recorded by Dr Pierre Barbet in his book *A Doctor at Calvary*. Dr Barbet writes of his personal conversations with the Pope on the subject[42].

Pope Paul VI also seemed to have a special veneration for the Shroud. In 1973 it was shown on television for the first time, with a filmed introduction by the Pope. Paul had also intended to visit Turin during the expositions of 1978; however he died before he could make this visit. The short-lived Pope John Paul I was also said to have been planning a visit before the close of the expositions. Despite the controversy over the radiocarbon dating of the Shroud, Pope John Paul II proceeded to authorise expositions in 1998 and 2000.

One could almost speculate that there is some information on the Shroud's origins in the Vatican that has never been made public but which has led popes over the centuries to recognize it as a genuine relic. It is however very easy to ascribe all sorts of conspiracy theories or secrets to the Vatican – many writers do it almost as a matter of course. Let it merely be said that the Shroud has attracted the attention and veneration of numerous popes and other Church authorities over the centuries.

Chapter 8 –
The Missing Years: Cathars or Templars?

How did Geoffrey de Charny come into possession of the Shroud that he exhibited in his church in Lirey? This is one of the enduring mysteries of our time and until and unless some clear, unambiguous evidence is found, it will always remain a matter of speculation. The approach to this question must be a scientific one.

There are no more than three alternative explanations for Geoffrey de Charny's Shroud: it was a forgery, crafted by some means not long before he obtained it; it was a genuine relic previously unknown to history; or it was a genuine relic known to history.

The idea that Geoffrey de Charny came into possession of a relic that had been unknown to history for some thirteen centuries can be discounted as extremely unlikely. This leaves the alternatives of a forgery or a known relic. There was only one relic known to history that resembled de Charny's Shroud; that was the Cloth of Edessa that disappeared from Constantinople in 1204.

That leaves two alternative questions to be answered. If de Charny's Shroud was a forgery, where did he obtain it from and how was it made? If de Charny's Shroud was the Cloth of Edessa, where did he obtain it and where was it from the time it disappeared in 1204 until it appeared in Lirey in about 1355?

The idea that the Shroud of Turin might be the actual burial cloth of Jesus Christ, imprinted with the marks of his Passion and therefore possible physical evidence of his Resurrection, is a very uncomfortable and very challenging one for many people. Acceptance of this idea would inevitably lead to consideration of the truth of the Christian religion and its teachings on moral issues in particular. It would also lead to consideration of the supernatural in general – an uncomfortable area for those who consider that they have their feet firmly planted on ground which is totally natural. Scientists, sceptics, agnostics, atheists, relaxed and relapsed Christians, all of these find it intellectually easier to accept a forged Shroud rather than a genuine one.

The radiocarbon dating of samples from the Shroud which took place in 1988 was, metaphorically, manna from Heaven for the sceptics. Three samples of cloth from the Shroud had been tested at three separate and highly reputable laboratories. They all agreed that the Shroud's raw flax had most likely been made into linen in about the year 1325 AD, give or take 65 years either way[1]. Radiocarbon dating is a tried, tested and thoroughly reputable scientific procedure. If these tests showed that the Shroud dated from the fourteenth century, what further proof could be needed? Indeed, the results of these tests were supported by the contemporary allegations of Bishop d'Arcis that Geoffrey de Charny's Shroud was a fraud, having been "cunningly painted". It was easy for sceptics to ignore other scientific studies of the Shroud and to cite this as "proof" that the Shroud is a mediaeval forgery.

There have been numerous theories as to how the Shroud was forged in the fourteenth century. Some of these will be considered in detail in Chapter 18. Suffice it to say at this stage that not one of them has been able to explain within the rigours of scientific proof how the forging was done, or to replicate it. At this stage I would rather consider attitudes towards relics in the thirteenth and fourteenth centuries and whether a blatant forgery would have had any public acceptability at all.

Professor Edward Hall of the Research Laboratory for Archaeology and the History of Art at Oxford University, one of the laboratories that carried out the testing on the Shroud samples, expressed his view as follows in a report in *The Independent* in London on 14 October 1988:

"There was a multi-million pound business in making forgeries during the fourteenth century. Someone just got a bit of linen, faked it up and flogged it.[2]"

He subsequently told an audience of the British Museum Society in London that radiocarbon dating had so conclusively proved the Shroud to be a fake that anyone who continued to believe it to be genuine had to be a "Flat Earther"[3].

In considering mediaeval attitudes towards relics it is necessary to distinguish between credibility and gullibility. There were many so-called relics at the time that today we would laugh at but which were genuinely believed in at the time. The greatest collection of relics in the Christian world before 1204 was in the Imperial Pharos Chapel in Constantinople. After the Latin conquest of Constantinople, the Latin Emperor Baldwin II between 1239 and 1242 sent a group of 22 relics from this collection to his relative Louis IX, the King of France. Louis IX had the Sainte Chapelle of Paris built to house this collection, which included the Crown of Thorns. Unfortunately the collection is no longer in existence as the Sainte Chapelle was plundered at the time of the French Revolution and most of its contents destroyed. These relics also included such fanciful items as nappies of the infant Jesus, milk of the Virgin Mary and the towel used by Jesus to dry the Apostles' feet[4].

The point is not that we would consider such items preposterous today but rather that they were fully accepted and believed to be genuine at that time. Indeed, not only were they accepted as genuine, they were treasured by Emperors and Kings and kept in the most ornate surroundings. These relics had credibility. What their actual origin was is anybody's guess, but the Byzantine Emperors and the Kings of France believed them to be genuine and treated them as such.

That there was a trade in fraudulent relics has been documented since early Christian times. The writings of St Gregory the Great and St Gregory of Tours in the sixth century refer clearly to unprincipled persons enriching themselves by a trade in relics, the majority of which were fraudulent[5]. This was obviously a matter of ongoing concern to the Church authorities as two Councils of Lyons, in 1245 and 1275, prohibited the veneration of recently found relics unless they were first approved by the Pope. This was emphasized by Bishop Quivil of Exeter in England in 1287 when he wrote:

The Missing Years: Cathars or Templars?

"We command the prohibition to be carefully observed by all, and decree that no person shall expose relics for sale, and that neither stones, nor fountains, trees, wood or garments shall in any way be venerated on account of dreams or on fictitious grounds.[6]"

This prohibition by the Church authorities would clearly have been of relevance to the exposition of the Shroud by Geoffrey de Charny in 1355.

There would seem to have been two levels of belief in relics. At the level of rulers, church dignitaries and the upper classes there was acceptance as being genuine of relics that would not be accepted today. Belief systems were different at that time. Believers then could accept things that are no longer credible in terms of our science and concepts. That is not to say that these mediaeval believers were gullible – they were not. They felt that they had reason to believe that certain artifacts were genuine and they had no opposing reason to deny their authenticity. These artifacts were credible.

At the level of the common people there was undoubtedly gullibility. Here was the market place for fraudulent relics. Unscrupulous traders would always have been looking for the opportunity to make a quick buck, just as they do today. Undoubtedly Professor Hall was on the right track when he suggested that there was a large trade in fraudulent relics during the fourteenth century.

But was Geoffrey de Charny one of the gullible or did he value what was credible? To put it another way, was Geoffrey de Charny, the paladin and statesman noted for his love of chivalry and honour, dishonest enough to be a conspirator in the forging of the Shroud or gullible enough to have been deceived by a trader in fake religious relics? Taking his character into account there can be no doubt that he genuinely believed that the Shroud in his possession was the genuine burial cloth of Jesus Christ. That he claimed it was is testified to by Bishop d'Arcis, who wrote in his supposed Memorandum to the Pope that "…it (the Shroud) was on the previous occasion (of its exposition) declared to be the true shroud of Christ.[7]"

This leads on to the question, where then did Geoffrey de Charny obtain the Shroud? Wherever it was it must have been a source that was credible to him and which gave him every cause to believe that the Shroud was the genuine article.

* * * * * * * * *

Several theories have been proposed regarding the whereabouts of the Shroud before its exposition in Lirey by Geoffrey de Charny. One of these involves the Cathars in the south of France and another particularly involves the Knights of the Temple.

The Cathar sect of southern France has been the subject of much speculative as well as factual writing. The brutality and bloodshed with which the Cathars were suppressed by the Catholic Church during the Albigensian Crusade of the early thirteenth century have left them with an aura of romantic tragedy.

The Cathars were dualists. This was a belief, considered heretical by the Church, which dated back to the early days of Christianity. The dualist belief was that all matter was the creation of the Demiurge, a spirit of evil who had made man in his own likeness. The true God was the God of love and, in dualist Christian belief, Jesus Christ was his messenger. The material world was therefore intrinsically evil and therefore to be avoided as much as possible. One of the earliest Christian dualists was Marcion in the middle of the second century. For baptized Marcionites marriage and sexual intercourse were strictly forbidden. The austerity of this belief was such that many dualists postponed baptism until late in life. As a result church organization divided believers into two classes – practitioners and believers, the two being separated by a ceremony of initiation. The knowledge of the initiated practitioner was known as *gnosis*, from which the term "Gnostic" is derived to also describe such sects. *Gnosis* is distinguished from *pistis*, the faith of the ordinary believer[8].

Marcion taught that Christ could not have assumed a material body; a material body and a physical birth belonged to the evil Creator and would be unworthy of the true Christ. Christ was revealed full-grown at once. Marcion accepted that Christ was crucified, but in a non-material body. Other Gnostics denied the Passion and death of Christ as well as his birth[9].

The Missing Years: Cathars or Templars?

The Marcionite heresy did not last long in its original form. Marcion had formed his sect in Rome, but following its fading away Gnosticism or dualism became primarily an eastern belief. It manifested itself again in Manichaeanism, the teaching of a Persian named Mani in the third century. This survived as a major religion in western Asia for six hundred years. Stricter Manichaeans avoided all contact with the material world, refusing to work, fight or marry.

In the tenth century Gnosticism returned to eastern Europe through an Armenian sect known as Paulicians. The first major dualist church of Europe was the Bogomils, who originated in Bulgaria and combined strict dualist beliefs with Slav nationalism. They refused to venerate the cross as it was both material and a symbol of Christ's death. They abstained from meat and wine and also from sexual intercourse[10].

With the start of the Crusades at the end of the eleventh century, western Europeans came into contact with dualist sects in the Balkans, in Asia Minor and in Palestine. By the middle of the twelfth century there were well-organised dualist churches in Cologne and Flanders. The Cologne dualists shared eastern dualist traditions and divided themselves into believers and initiates. The latter were called Cathars from a Greek word meaning "purified". Persecution of these sects by the Church led to a large number of them fleeing to the south of France, in the region known as Languedoc, an area where the civil power was weak and the nobility did little to co-operate with the Church. Here Catharism took root and flourished[11].

One hypothesis proposed regarding the Shroud's whereabouts before its arrival in Lirey is that it was in the possession of the Cathars of Languedoc[12]. According to this hypothesis the Cloth of Edessa was deliberately stolen by dualists in Constantinople during the chaos of the sack of the city by the Fourth Crusade. These Greek dualists sent it to their fellow-believers in Languedoc for use as a palladium – an article which provides miraculous protection against hostile forces – against the increasingly hostile Catholic Church[13].

The Cloth of Edessa had the reputation of having such mystical powers. It had reputedly saved the city of Edessa from a siege by the Persian army in 544 AD and, according to the French crusader Robert de Clari, "the Byzantine Emperor had always relied on his relics to protect his throne and his city."[14]

The situation in Languedoc in the latter part of the twelfth century was typified in two ways – the increasing influence of heretical churches, particularly the Cathars, and a complicated system of continually-changing feudal allegiances. The most powerful noble in the region was the Count of Toulouse. Two events set in motion the train of events that led to the Albigensian Crusade. In 1194 Count Raymond V of Toulouse died, to be succeeded by his son Raymond VI, a much weaker and less decisive individual with sympathies towards the Cathars, and in 1198 Innocent III was elected as Pope. Innocent III was a lawyer by training with strong views on the unity of the Church[15]. His hatred of heresy was deeply felt and strongly expressed. The Pope and the Count were to be the protagonists of the Albigensian Crusade.

It was in 1207 that Pope Innocent publicly proclaimed a military campaign against the heretics of Languedoc. Up to that time he had relied primarily on papal legates to try and impose his authority on the local nobility and hence on the populace. It is therefore unlikely that the Cathars would have felt so in need of protection against a military threat in 1204 that their co-religionists in Constantinople would have gone to the extent of stealing the Cloth of Edessa to send it to the south of France.

The Albigensian Crusade was launched in 1208, following the murder of a papal legate possibly at the instigation of the Count of Toulouse. The objective of the Crusade was to restore papal authority in the south of France and to destroy the Cathar heresy; it consisted of a series of military campaigns that lasted until 1229 and included the massacres of civilians and sacks of cities that were an integral part of mediaeval warfare. It resulted in the consolidation of the rule of the King of France in the Languedoc region but the Cathar heresy continued to survive. Its destruction then became the responsibility of the Inquisition[16].

After the conclusion of the Crusade the Cathars maintained control of the hill-top fortress of Montsegur, near Foix. In 1242 the commander of Montsegur led a small force to kill four Inquisitors in a nearby town. The result of this was the launching of an attack on Montsegur by a French army. Montsegur surrendered in early 1243. While the military garrison escaped with their lives, more than two hundred captured Cathars were burned to death in the plain below the castle[17]. After this the Cathars ceased to be an organized church with a doctrine and hierarchy. The Inquisition suspended its activity in Toulouse in 1279. By the end of the thirteenth century the surviving Cathars were scattered and disorganized. The last Cathar Perfect to fall into the hands of the Inquisition was William Belibaste who was burned at the stake in 1321[18]. However, surviving groups of Cathars may have existed until about 1350, when the Black Death had killed some 50% of the population of Languedoc.

The Cathar theory of "the Missing Years" is that the Cloth of Edessa was sent to Languedoc in 1204 and possibly taken to the newly restored fortress of Montsegur. After the fall of Montsegur in 1243 four Cathars are said to have escaped from the fortress with a mysterious treasure that had been rumoured to be held there. This treasure may have been, or included, the Cloth of Edessa[19], which remained in the possession of surviving Cathar families until they were all wiped out by the Black Death. It is fact that in 1349 Geoffrey de Charny was granted by the King of France an annuity payable from the first forfeitures which might occur in the Languedoc towns of Toulouse, Beaucaire and Carcassonne. It could be speculated that de Charny acquired title to the Cloth of Edessa through such a forfeiture, possibly in Carcassonne[20], and that this Cloth then became the Shroud of Lirey.

This hypothesis does answer the key questions: how was the Shroud taken from Constantinople, how did it come into the hands of Geoffrey de Charny and what is the link between these two events? However there is not one iota of historical evidence or even reliable hearsay to support it.

It has also been argued that the Cathars had a fear or horror of the crucifixion. This does not accord with their dualist beliefs which professed that Jesus had merely inhabited a representation of a human body and had therefore not been a physical being. In terms of such beliefs the Crucifixion would have been only a mirage. But it has been argued that this fear or horror, together with other apparent views they had on physical aspects of the Crucifixion, is evidence that the Cathars had obtained detailed information about the Crucifixion and that this knowledge could only have come from their possession of the Shroud[21].

Mark Oxley

* * * * * * * * *

Much has been written about the Knights Templar, some factual, some speculative and some pure fiction. The Order of Poor Fellow-Soldiers of Christ and the Temple of Solomon was founded in 1118 as an order of monastic knights who had taken a vow of poverty, chastity and obedience. They took their name from the quarters assigned to them by King Baldwin II of Jerusalem – a portion of the al-Aqsa Mosque on the Temple Mount, where the original Jewish Temple had once stood. Nine years later the founder of the Order, Hugh de Payens, traveled from the Holy Land to Rome to seek official papal sanction for his order. He had obtained strong support from the Abbot of Clairvaux, Bernard, who was at that time one of the most influential churchmen in Europe and who was also, coincidentally, a cousin of Hugh. Bernard was a reformer who wanted to purify the Church and to destroy its enemies; he was filled with enthusiasm at the concept of an order of knights functioning under monastic vows. He assisted in shaping the Order and defined its aims and ideals in a Rule governing its conduct.

Bernard also called for gifts of land and money for the Templars. The Count of Champagne was the first to respond and gave the Templars a grant of land at Troyes. This became the Templars' first "preceptory", a term used for establishments that acted as supply bases to support Templar operations in the Holy Land. The Templars received numerous further grants of land in France, Normandy, England and Spain. The officers in charge, the preceptors, were charged with extracting the maximum revenue from these properties. This revenue was used to supply weaponry and equipment to the Templars and was undoubtedly the foundation of the Order's later wealth.

The Templars were originally founded to offer protection to pilgrims traveling to Jerusalem. They rapidly became a standing army that fought in defence of the Crusader states in the Holy Land. Pope Innocent II issued a bull in the first half of the twelfth century exempting the Templars from all authority on earth except that of the Pope himself. As well as fighting the Templars became involved in financial services and developed some of the systems of modern banking. They introduced revolutionary ideas into warfare as well. The Templar rule called for full and immediate obedience of the knight to his superiors; he was therefore trained to move quickly in

The Missing Years: Cathars or Templars?

battle without question and this developed into formal drill techniques, the first to be used in European warfare since the time of the Roman legions[22].

As a fighting force the Templars were noted for their ferocity and their courage. At times their belief in their own invincibility worked against them. In 1153 the Templars, under the leadership of their Grand Master, Bernard de Tremelai, were involved in a siege of the Egyptian-held city of Ascalon. The besiegers forced a breach in the city wall; the Grand Master led a force of forty Templars into the breach and then ordered the remainder of the Templars to prevent any other Christian forces from following him into the city. It was as if he had decided that forty Templars by themselves could defeat thousands of well-armed Egyptian troops and capture the city. The Grand Master and his small force were inevitably annihilated[23].

The Templars served in the Holy Land for nearly two hundred years. By the late thirteenth century the Crusader states had been virtually destroyed by Muslim armies. In 1291 the Templars abandoned their last two castles in the Holy Land, bringing the Crusades to an end[24]. Shortly after this the Templars elected a new Grand Master, Jacques de Molay. An illiterate man, he was a firm believer in military discipline whose primary ambition was to serve in another Crusade to recapture Jerusalem.

At this time the King of France was Philip IV, known as "the Fair" (for his good looks, not his sense of justice). He had been at war with King Edward I of England for all his adult life and to pursue this war he had taxed, borrowed and confiscated all the money he could. He had borrowed heavily from the Templars. By the beginning of the fourteenth century it had become clear to Philip that one way of resolving his financial difficulties would be by getting his hands on the accumulated wealth of the Templars in France. In 1305 Philip was instrumental in having a Frenchman elected as Pope, who took the papal name of Clement V[25]. This Pope never took up residence in Rome, but stayed at various locations in France before purchasing Avignon in Provence as the new seat of the papacy.

Clement V summoned Jacques de Molay to the papal court which was then at Poitiers, ostensibly to discuss a new Crusade. The reality was that Philip IV had decided on the suppression of the Templars throughout Europe, so that he could seize their assets in France without risk of retribution from Templars in other countries. To do this he needed the co-operation of the Pope and he needed to have the Grand Master in his hands. Philip carried out a dress rehearsal of the suppression of the Templars in 1306 when he ordered the arrest and imprisonment of all the Jews in France and the confiscation of all their records and assets. All money owed to the Jews was ordered to be paid to the King of France[26].

The Templars were exempt from all secular law and answerable only to the Pope. For this reason Philip decided that their suppression would have to be based on heresy and offences against God, and that these offences would have to be offences of the Order as a whole rather than the offences of individual Templars. Only in this way could the confiscation of the Order's property be justified. Philip IV had very clever legal advisers. At the same time Philip, through his counselor William de Nogaret, started the process of blackening the Templars' name through allegations of sorcery, homosexuality and heresy[27].

Jacques de Molay returned to France in early 1307 and marched from Marseilles to Paris accompanied by a personal escort of sixty knights and a large amount of gold coins. In Paris he was informed of rumours regarding sinful and unholy conduct within the Templars. He then traveled to Poitiers to seek a formal papal enquiry into these rumours. He had no idea of the machinations of Philip IV and the assistance the latter would be given by Pope Clement V. De Molay was a simple, unsubtle soldier who could never have imagined his King and his Pope of being guilty of the treachery that was in hand.

Philip now presented the Pope with "evidence" he had obtained on Templar corruption. Among the allegations made were that new members, at their initiation, were required to spit or trample on the cross and to deny Jesus Christ. They desecrated the holy sacraments. They worshipped Satanic idols in the forms of a bearded head and a cat[28]. The Pope agreed to hold a formal enquiry into the charges. Philip forestalled him by ordering the arrest of all the Templars in France, which took place on Friday 13th October 1307, a date that has been considered unlucky ever since. Despite initial protests at this flouting of papal authority by the King of France, Clement eventually gave his full approval to Philip's actions.

The most horrendous tortures were used to obtain confessions from the Templars of their heresy and corruption. At one enquiry a footless Templar was carried in carrying a bag of blackened bones that had dropped out of his feet as they had been burned off. The Templar treasurer was reported to have said, "Under such torture I would willingly confess to having killed God." Templars confessed to worshipping a head, or a head with three faces, or a head with four feet, or a head with two feet. It was a metal head, or a wooden head, or a human skull. Some Templars confessed that the head was named Baphomet.

The papal commission of enquiry was convened nearly two years after the arrest of the Templars. All the Templars who appeared before the commission except one denied the truth of the confessions that had been forced from them. When the commission sought an individual to act as defender of the Order, five hundred and forty-six Templar knights in Paris volunteered. Eventually it became clear to Philip IV and his advisers that the enquiry was not going well from their point of view. Again the Pope was pre-empted. Philip arranged for the brother of one of his ministers, Philip de Marigny, to be appointed to the vacant position of Archbishop of Sens. De Marigny rapidly called a council of his bishops to condemn the Templars. As a result of his efforts fifty-four Templars who had renounced their forced confessions were sentenced to be burned at the stake. Not one of these men recanted to save his life. De Marigny's actions were, however, sufficient to intimidate the remaining witnesses and the commissioners of the papal enquiry. Eventually in 1312 the Pope conceded to the King of France and announced the suppression of the Order. Philip already had all the Templars' assets in France, at least some of their treasure and freedom from his debts to the Templars[29].

The final act came in 1314 when de Molay and three of his officers were brought out before the Cathedral of Notre Dame to confess their crimes and sins. De Molay took this last opportunity to declare his innocence and that of his Order, with the result that he was burned to death at the stake that same evening. According to legend de Molay cursed Philip and Clement from the stake and summoned them to meet him before the throne of God within the year to answer for their crimes. Both men died during that same year.

De Molay was accompanied to his death by the Templar Preceptor of Normandy, a knight by the name of Geoffrey de Charnay[30].

Much has been made of the similarity in names between the Preceptor de Charnay and Geoffrey de Charny of Lirey. It has been suggested that they were related and that this family link resulted in the Lirey de Charny obtaining the Shroud from the Templars[31]. In fact the only evidence for the Templars ever having had possession of the Shroud is this similarity of names and the allegations that the Templars worshipped a bearded head.

In fact it is likely that the similarity of names between the Preceptor of Normandy and the de Charny of Lirey is one of those coincidences that litter history. The Preceptor of Normandy came from a family in Anjou, west of Paris[32]. Geoffrey de Charny of Lirey was from the Champagne region, north-east of Paris. A lot of research has been done in recent years on Geoffrey de Charny's family and his ancestry. While it has become clear that he was descended both from the Latin Emperors of Constantinople and from King Louis IX's close companion, Jean de Joinville, no connection has been found to any de Charnay family in Anjou[33].

If it is assumed that the Shroud of Lirey and the Cloth of Edessa lost at Constantinople are one and the same, it has to be noted that there were no Templars at the sack of Constantinople, despite suggestions to the contrary by a number of writers. The French knight Robert de Clari, in his book *The Conquest of Constantinople,* lists in detail the leading members of the Fourth Crusade. He makes no mention of Templars[34]. It has to be borne in mind that the Templars served as a standing army in the Holy Land and their recruits traveled directly from Europe to Palestine to join them. The major Crusades, such as the Fourth Crusade, were made up of feudal landowners, their retinues, feudal subordinates and men-at-arms in general from western Europe. It has, however, also been suggested that the Templars may have purchased the missing Cloth of Edessa from some anonymous Crusader who had acquired it during the Sack[35].

The connection between the Templars and a bearded head *may* have been one of the fabrications devised to bring about their downfall in the early 1300s. There does not seem to be any firm evidence of any such connection before this time. All contemporary references to the head seem to have been drawn from confessions made by Templars to their inquisitors

The Missing Years: From Athens to France

in France after 1307. However, despite this, some of the references made to the head by imprisoned Templars seem to have a ring of authenticity. One Templar, Stephen of Troyes, stated that he had seen a mysterious image at one of the Templars' Paris ceremonies:

"…brought in by the priest in a procession of the brethren with lights; it was laid on the altar; it was a human head without any silver or gold, very pale and discoloured, with a grizzled beard like a Templar's.[36]"

There is also a reference in the *Chronicles of St Denis* describing this cloth as seeming like "…an old piece of skin, as though all embalmed and like polished cloth.[37]"

While the other allegations against the Templars included such clearly spurious charges as sorcery and demon-worship, the allegations about the head could have been based on actual veneration by the Templars of a genuine Christian relic. The existence of this head, therefore, would not in itself necessarily support the charges of sorcery and heresy leveled by Philip IV. Indeed false charges may be given more credibility when they are derived from a factual but innocent basis. The question remains, however, if the Templars did indeed have some image of a head, was it the Cloth of Edessa? This question will be addressed in detail in the next Chapter.

The difficulty remains of providing a plausible explanation of how such a venerated relic came to fall into the hands of a relatively modest family in the Champagne region, even if that family had been related to a leading Templar. If the Shroud had been part of the so-called Templar treasure in Paris, it would either have fallen into the hands of the King and his counselors or would have been part of the treasure that, according to legend, was smuggled out of France by sea at the time of the arrest of the Templars and that was possibly taken to Scotland. In neither case would it have appeared in Lirey.

Chapter 9 –
The Missing Years: From Athens to France

There is another possible explanation of how the Cloth of Edessa was taken from Constantinople and eventually came into the possession of Geoffrey de Charny. This explanation has the advantage of simplicity and avoids the involvement of organizations as prone to controversy and sensationalism as the Templars and Cathars. It relies on two pieces of factual evidence, one dating from shortly after the sack of Constantinople and one dating from the time of Geoffrey de Charny's first exposition of the Shroud at Lirey.

Firstly, it is necessary to understand the effect that the Fourth Crusade had on the Byzantine Empire. Following the capture of Constantinople by the Crusaders in 1204, the Byzantine Empire literally shattered. There were no less than seven successor states – three Latin or Crusader states and four Greek states. In addition Venice took possession of some of the former imperial territory[1].

One of the Greek successor states was Epirus on the west coast of Greece. Following the fall of Constantinople this came under the rule of Michael Ducas, an illegitimate cousin of the Byzantine Emperor Isaac II Angelus, who had contributed so much to the downfall of Constantinople. Michael had a legitimate brother, Theodore Angelus Ducas, who succeeded him in due course as Despot of Epirus and who eventually proclaimed himself Emperor after capturing the Latin Kingdom of Thessalonica adjacent to Epirus in northern Greece. Theodore Ducas clearly saw himself as a legitimate claimant to the imperial throne of Byzantium and he was equally clearly concerned about what he considered to be imperial property.

On 1 August 1205 he wrote a letter to Pope Innocent III, complaining about the looting of Constantinople by the Fourth Crusade. In his letter he said:

"In April last year a crusading army, having falsely set out to liberate the Holy Land, instead laid waste the city of Constantinople. During the

The Missing Years: From Athens to France

sack troops of Venice and France looted even the holy sanctuaries. The Venetians partitioned the treasures of gold, silver and ivory, while the French did the same with the relics of the saints and, most sacred of all the linen in which our Lord Jesus Christ was wrapped after his death and before the resurrection. We know that the sacred objects are preserved by their predators in Venice, in France and in other places, *the sacred linen in Athens* (author's italics). So many spoils and sacred objects should not be taken contrary to all human and divine laws, nevertheless in your name and in the name of Jesus Christ our Lord, albeit against your will, the barbarians of our age have done just that.

"The teaching of Jesus Christ our Saviour does not allow Christians to despoil other Christians of their sacred belongings. Let the thieves have the gold and silver, but let what is sacred be returned to us; and toward this end my brother and lord has placed his greatest trust in the intervention of your authority. With your co-operation, restitution is certain. The faithful await your action and your co-operation. My brother and lord, Michael, awaits the justice of Peter." [2]

This is a strongly worded letter and the strength of Theodore's feelings can be judged by the fact that he had taken the trouble to journey from Greece to Rome, which is where the letter was written.

Theodore Angelus Ducas's claim that the Cloth of Edessa was in Athens may be supported by a letter written in 1207 by Nicholas of Otranto, the Abbot of Casole monastery in southern Italy. In 1205 Nicholas met with the new Papal legate, Benedict of St Susanna, who was then on his way to Constantinople. Nicholas, who was fluent in both Latin and Greek, accompanied Benedict through Greece to Constantinople, where he served as Benedict's personal interpreter and translator.

Nicholas made an apparent reference to the shroud of Jesus in the midst of a discussion in 1207. As part of this discussion Nicholas referred to relics of the Passion, including a portion of the bread from the Last Supper and Jesus's *spargana*, a Greek word meaning linens, which must be interpreted as burial linens in the context of the Passion. Nicholas wrote:

"When the city was captured by the French knights, entering as thieves, even in the treasury of the Great Place where the holy objects were placed, they found among other things the precious wood, the crown of thorns, the sandals of the Saviour, the nail, and the *spargana* which we (later) saw with our own eyes."

He does not make clear where he had seen the *spargana*. However, in 1206 he and Benedict had traveled through Thessalonica and Athens, debating questions of Church unification with Greek theologians. The reference to seeing the linens "with our own eyes" is a peculiar remark which would seem out of place if he had seen them in Constantinople. The inference is that he had seen them on his travels to Constantinople, which were of course subsequent to the sack of that city[3].

There had been in Constantinople several linens claimed to be the funeral linens of Jesus. Most of these had been kept in the Pharos Chapel in the Imperial Palace, together with other relics. It was from this collection that the Emperor Baldwin II had sent to King Louis IX of France a group of relics, said to include a piece of the funeral cloths. However there is no mention anywhere of there being an image on these cloths and the material of these cloths has since been shown to be distinctly different from that of the Shroud of Turin[4].

As we have already seen, the Cloth of Edessa was not in the Pharos Chapel at the time of the Fourth Crusade, but rather in the Church of St Mary in Blachernae, as described by Robert de Clari:

"There was another of the churches which they called My Lady Saint Mary of Blachernae, where was kept the *sydoine* (shroud) in which Our Lord had been wrapped, which stood up straight every Friday so that the features of Our Lord could be plainly seen there."[5]

(This is not an attempt to describe a miraculous happening. Undoubtedly some mechanical contraption was involved in the raising of the Shroud, as had no doubt been the case in the Edessa Easter rituals described in Chapter 4.)

This is a clear reference to a shroud with an image on it and it must be concluded that this was indeed the Cloth of Edessa, rather than any relic in the Imperial Palace. De Clari made it quite clear as he continued his description of the shroud that it disappeared at the time of the sack, and that it did not remain in Constantinople:

"And no one, either Greek or French, ever knew what became of this *sydoine* after the city was taken."[6]

Its weekly exhibition no doubt brought it a lot of publicity and hence made it a perfect target for the looting crusaders.

There is a graphic description of the sack of Constantinople in Sir Steven Runciman's *A History of the Crusades*:

"The sack of Constantinople is unparalleled in history. For nine centuries the great city had been the capital of Christian civilization. It was filled with works of art that had survived from ancient Greece and with the masterpieces of its own exquisite craftsmen. The Venetians indeed knew the value of these things. Wherever they could they seized treasures and carried them off to adorn the squares and churches and palaces of their town. But the Frenchmen and Flemings were filled with a lust for destruction. They rushed in a howling mob down the streets and through the houses, snatching up everything that glittered and destroying whatever they could not carry, pausing only to murder or to rape, or to break open the wine cellars for their refreshment. In St Sophia itself drunken soldiers could be seen tearing down the silken hangings and pulling the great iconostasis to pieces, while sacred books and icons were trampled under foot. While they drank merrily from the altar-vessels a prostitute set herself on the Patriarch's throne and began to sing a ribald French song. Nuns were ravished in their convents. Palaces and hovels alike were entered and wrecked. Wounded women and children lay dying in the streets. For three days the ghastly scenes of pillage and bloodshed continued, till the huge and beautiful city was a shambles. Even the Saracens would have been more merciful, cried the historian Nicetas, and with truth[7]."

It can be easily imagined that in such chaos a group of Crusaders could have taken the opportunity to snatch a prominent relic that was not protected in the Imperial Palace.

Theodore Angelus's letter is of great importance to the history of the Shroud of Turin. It is the only firm evidence immediately following the sack as to the whereabouts of the Cloth of Edessa.

The Duchy of Athens was one of the Crusader successor states to the Byzantine Empire. The first Duke of Athens was Otto (or Otho) de la Roche, a Burgundian lord who took part in the Fourth Crusade. The Duchy of Athens was originally a vassal state of the Kingdom of Thessalonica in Greece, another of the Crusader successor states. This kingdom had been founded by the leader of the Crusade, Boniface Margrave of Montferrat, who assumed the title of King. It would seem likely that Otto de la Roche received Athens as a reward from Boniface for his services during the Crusade. This in turn would suggest that Otto had been prominent during the Crusade and therefore in a position to acquire valuable loot during the sack of Constantinople.

Otto's palace in Athens consisted of the buildings of the Acropolis, including the Parthenon, which was not in its present ruined state at that time. If the Shroud was indeed in Athens, this is probably where it would have been kept.

In 1225 Otto handed over his lordship of Athens and all his Greek lands to his son Guy and returned to Burgundy with his wife and two other children of theirs[8]. (Although Otto had styled himself "Duke", the lordship of Athens only formally became a Duchy in 1260.) Guy remained in Athens as its ruler and the de la Roche family ruled there until 1308, when the last de la Roche duke was succeeded by his nephew Walter de Brienne. At this time there was a mercenary band in the region fighting for the Byzantine Emperors, who had since been restored to Constantinople. This band originated from Aragon in Spain and was known as the Catalan Company. Following a mutiny they arrived in the vicinity of Athens where they were hired by Duke Walter in 1310 to fight on his behalf against the Greek states of Epirus and Nicaea. They repaid him the following year by deposing him and taking over the Duchy.

There are two suggestions as to how the Shroud got to France from Athens. The most obvious one is that Otto de la Roche took it with him when he returned home. It is highly unlikely that he would have left such a valuable piece of loot in Greece when he could easily take it back to his home in Burgundy. However, an alternative theory was proposed by Monsignor Arthur Barnes, the Domestic Prelate to Pope Pius XI, in 1934. He suggested that Otto de la Roche sent it, when the opportunity offered, to his family in Franche-Comte, which was then part of the County (as opposed to Duchy) of Burgundy[9].

It is not inconceivable that Otto de la Roche sent the Shroud back to France before he himself left Athens. This could well have been at a time when he felt threatened in his rule of Athens and felt that it would be prudent to send his most valuable possessions to a place of safety. Such a time could have been immediately prior to 1224, when the Kingdom of Thessalonica, the feudal overlord of the Duchy of Athens, was captured by Theodore Angelus Ducas, the Despot of Epirus, who subsequently assumed the title of Emperor. In view of his complaint to the Pope it might have been expected that he would cast an eye on Athens with a view to recovering imperial property he had reason to believe was there. However Theodore decided instead to campaign against the Bulgarians on his northern border, where he was defeated and taken into captivity. It was shortly after this that Otto himself returned home.

Otto remained in France until his death in 1234. From his first wife, Isabelle de Ray, he had obtained the title of Baron of Ray. This title descended to his son, Otto II, and then to his grandson Jean de Ray. Otto II had a daughter, Elizabeth. She married Henri de Vergy, a member of another leading Burgundian family. Their son, Jean I de Vergy, eventually became Seneschal of Burgundy. He had three sons, one of whom, Guillaume, Lord of Mirabeau, was the father of Jeanne de Vergy, second wife of Geoffrey de Charny[10]. Jeanne was therefore descended in the fifth generation from Otto de la Roche.

At Ray-sur-Saone in Burgundy there is a chateau which is still owned by a descendant of Otto de la Roche. In the museum at the castle there is a fragment of the True Cross which, according to tradition, Otto brought from Constantinople. Also at the castle is a chest which, again according to family tradition, was used to transport the Shroud from Athens to France[11]. The label on the chest reads:

"13th century coffer in which was preserved at Ray Castle the Shroud of Christ brought by Otho de Ray from Constantinople. 1206"

Mark Oxley

The label is modern and merely reiterates the family tradition. Family memoirs record Otto as returning to his castle in France in 1206, bringing the Shroud with him. In fact his presence in Athens as late as February 1225 is recorded in a papal bull of Pope Honorius III[12].

The second piece of historical evidence in this chain is the Shroud medallion now kept in the Cluny Museum in Paris. This depicts the Shroud with its full body image, together with the coats-of-arms of both Geoffrey de Charny and Jeanne de Vergy. For some reason Jeanne de Vergy was specifically associated with the Shroud when it was exhibited in Lirey in 1355. One reason for this could well have been that it came from her family, through her direct descent from Otto de la Roche.

There are other family connections between the de Charny family, Greece and Constantinople, but the de la Roche/de Vergy connection seems to offer the best explanation of how the Shroud came from Constantinople to Lirey. It answers the key questions: what happened to the Cloth of Edessa in Constantinople; how did the Shroud come into the hands of Geoffrey de Charney; and what is the direct link between these two events?

It remains for a forensic analysis to be carried out of the interior of the chest in the castle at Ray-sur-Saone, to search for any traces of dirt, pollen, fabric or indeed human blood that might connect it to the Shroud of Turin. A superficial examination of the chest has been carried out by the Spanish Shroud researcher Cesar Barta. He and some companions had the opportunity to examine the chest and to illuminate it with ultraviolet light, which should have shown up any discernible traces of blood. Sr Barta has noted that they did not see any dust or human blood. There did seem to be some traces of textile fibre, of which he took a sample. This may, however, only be the residue from a sheet that was put inside the chest some years ago by a Mr Antoine Legrand, to see how the chest would match the dimensions of the Turin Shroud. Sr Barta was also told that the only original part of the chest is the bottom[13]. A detailed forensic examination may still bring to light evidence that would be missed by a superficial examination with the human eye.

* * * * * * * * *

It is at this point that it again becomes important to distinguish between the Cloth of Edessa and the Mandylion. The Mandylion also disappeared from Constantinople and may have left a trail that has led to confusion over the whereabouts of the Cloth of Edessa. It is only by accepting that the Mandylion and the Cloth of Edessa were two separate articles that a lot of this confusion can be cleared up.

To return to the sack of Constantinople, the Cloth of Edessa had been displayed in the Blachernae Palace while the Mandylion and the Keramion were kept in the Bucoleon Palace. These two palaces were about four miles apart. After the flight of Alexius V the senior remaining member of the Imperial family left in Constantinople was the Empress Maria, the widow of the deposed and blinded Isaac II Angelus. Maria was not Greek, but the daughter of King Bela III of Hungary and a French noblewoman, Agnes de Chatillon. Through her mother she was related to many of the Crusader leaders. At the time of the capture of Constantinople by the Crusaders she was in the Bucoleon Palace, which was surrendered to the leader of the Crusade, Boniface of Montferrat, on condition that the lives of all the people in the Palace were spared. The Bucoleon Place was spared during the sack of the city that followed. The fact that the Emperor Baldwin II was later able to send such a notable collection of relics to King Louis IX is a clear indicator that there was no looting of relics in the Bucoleon, which housed the main imperial collection of relics.

The Blachernae Palace surrendered to Count Henry of Flanders, but it may not have been given the same protection from the sack as the Bucoleon. In any event, as Robert de Clari recorded, the Cloth of Edessa disappeared from the Blachernae and neither French nor Greek knew where it had gone. He was quite emphatic on that point.

On 15 May 1204 the Empress Maria married Boniface of Montferrat. Within a few weeks Boniface left Constantinople to claim the Kingdom of Thessalonica that had been awarded to him. His new wife accompanied him. Boniface had been ambitious for the Imperial Crown and no doubt felt thwarted by its award to Count Baldwin of Flanders. It can be speculated that he and his wife took with them the Mandylion and the Keramion as compensation. Whatever happened, these two relics had been in the Bucoleon Palace under the protection of his wife, the then Empress Maria, and they were never seen again.

It may be significant that in the city of Salonika, which was the capital of Boniface's kingdom, there is an ancient church dating from the ninth or tenth century which was originally known as the Church of the Mother of God. Sometime between 1205 and 1222 its name was changed to the Church of the *Acheiropoietos* ("not made by human hand"), the term associated with the Cloth of Edessa and the early copies that had been made of it. This may be an indication that the Mandylion was kept in this church at some stage during the time that Maria spent in the kingdom.

Boniface was killed in battle in 1207. Maria married again and was soon widowed again. She remained in Thessalonica until 1224, when the Kingdom fell to Theodore Angelus Ducas, now the Despot of Epirus. Maria returned to Hungary. If she had the Mandylion she did not leave it in Greece, so it is probable that she took it to Hungary with her. Here there is another clue. There is a well-known icon, the Holy Face of Laon, that dates to the early thirteenth century. It bears a Slavonic inscription stating that "This is the Lord's face on the Cloth" and it may well be a direct copy of the Mandylion. It is believed to have been painted in northern Serbia, in an area which was part of Hungary during the thirteenth century. Whether there is any connection between the Empress Maria's travels and the painting of the Holy Face of Laon is a matter of pure conjecture, but there is a clear connection between the icon and the Mandylion. (The icon was sent to Laon in France in 1249 and has remained there since.)[14]

One can take this speculation further. Possibly the Mandylion somehow subsequently came into the possession of the Templars, possibly through some commercial transaction, and became the basis for the claims about that order revering a mysterious head[15]. Perhaps this was indeed more than the torture-induced raving of men in unspeakable agony. Where then could it have gone when the Templars were suppressed by Philip the Fair? According to testimony given to the Inquisition by the Templar John de Chalons of Poitiers, the Preceptor of the Order in France, Gerard de Villiers, knew in advance of the mass arrests of 13 October 1307 and fled from Paris with fifty knights, whom he commanded to put to sea on eighteen Templar galleys. Another knight, Hugh de Chalons, had fled with all the Templar treasure. Although Gerard de Villiers and Hugh de

Chalons did indeed escape the first arrests and were captured only several days later, anything they had had with them had gone[16].

This is the origin of the legends of the missing Templar treasure. Certainly if the Templars had had a particularly revered icon, it would have been sent to safety with any other treasure that was saved. One legend has it that the Templar treasure is buried in a vault under Rosslyn Chapel near Edinburgh in Scotland. Ground scans were carried out at Rosslyn Chapel in the early 1990s to try to detect any underground vaults, but no conclusion was reached. Since the establishment of the Rosslyn Chapel Trust in 1995 no attempts have been made to look beneath the Chapel, which is protected by various Scottish laws[17].

Curiously, in Rosslyn Chapel there is a carving of a figure holding up a cloth with an image of a head on it. The figure's head has been cut off so it is impossible to tell whether the figure represents a Templar[18]. The cloth in the carving appears to be more like a veil than a full-length shroud or even a folded shroud. Perhaps this is a representation of the Templar's venerated relic, the Mandylion. To extend this speculation even further, might it not be possible that the Mandylion, together with the Keramion, is buried under Rosslyn Chapel awaiting discovery at some future time?

However, the mystery does not end there. Ian Wilson, in *The Turin Shroud*, refers to a further curious incident where a panel painting of a bearded head, dated to about 1280, was found in an old Templar preceptory at Templecombe in Somerset, England[19]. There are striking resemblances between this painting and the Turin Shroud[20]. There is no evidence of the Shroud ever having been in England, so the unknown painter must have been painting from a memory or sketch of an image he had seen somewhere else. There are at least two possibilities. The first is that he had in fact seen the Shroud, but, as we have seen, it is unlikely that this had been in the possession of the Templars, although he may, quite by chance, have seen it somewhere else. The second is that he had seen a close copy of the Shroud. This also gives rise to two possibilities. One is that, as has been suggested, the Templars did indeed have the Mandylion, as opposed to the Cloth of Edessa, and that he had seen this. The other is that he had been to Rome in the Holy Year 1300 and had seen the exposition of the Veil of Veronica there. Pope Boniface VIII proclaimed the year 1300 to be a Holy Year and ordered weekly expositions of the Veil of Veronica. According to contemporary writers, the total number of visitors to Rome during that year reached two million. While this may be an exaggeration it leaves no doubt that pilgrims traveled to Rome from all over Europe for this event[21]. Templars would undoubtedly have been among the pilgrims. If either or both the Veil of Veronica and the Mandylion were themselves close copies of the Cloth of Edessa, and the Cloth of Edessa can be identified with the Shroud of Turin, this could explain how a talented artist at the end of the thirteenth century produced a portrait in England with marked similarities to the face on the Shroud of Turin[22].

* * * * * * * * *

One question that remains to be answered is why the Shroud, or the Cloth of Edessa if they are indeed identical, remained hidden for one hundred and fifty years. Many writers have suggested that legitimate owners of the Shroud would have exhibited it and that the fact that it was not exhibited indicates furtiveness, guilt or some degree of illegitimacy in its changes of ownership. The sudden appearance of the Shroud in Lirey, apparently from nowhere, is also used to support the theory that it is, after all, only a mediaeval forgery.

The truth may be quite simple. The Cloth of Edessa was one of the great relics of the Byzantine Empire. Theodore Angelus Ducas, in his letter to the Pope, described it as "the most sacred of all". The man who obtained possession of the Cloth of Edessa may have found himself in the position of a man who, today, steals the Mona Lisa. He has an extremely valuable article, the appearance of which is widely known. If it becomes known that he has it, the original owner will want it back very badly and someone else may wish to take it for himself.

There is no evidence of the Cloth of Edessa ever having been exhibited in Athens. Otto de la Roche and his successors probably had good reason to suspect that if they did show it off, it would attract the unfavourable attention of the Ducas family, firstly in Epirus and then, later, immediately to their north in Thessalonica. Thessalonica subsequently fell into the hands of the Byzantines of Nicaea, another successor state, who would have been just as interested as Theodore Angelus Ducas in recovering any stolen imperial property remaining in the region.

Once in France the Shroud would have been equally at risk. The Kings of France would not have hesitated to take possession of the Shroud had they known it was in their domain. Louis IX has already been shown to have a great interest in relics, having built the Sainte-Chapelle in Paris to house his collection. Undoubtedly he would not have hesitated to add "the most sacred of all" to his collection had he known its whereabouts. His two successors, Philip III and Philip IV, would have been equally unhesitating. Philip IV's avarice has already been demonstrated in the way he suppressed the Templars in order to escape his debts and take over their assets.

To complicate the issue, there was also the need to obtain approval from the Pope before relics could be exposed or venerated. The Fourth Lateran Council which started in 1215 had banned the transaction of relics, condemning it as sacrilegious. This would not necessarily have been so important in Athens, away from the immediate influence of the Pope, but would have been significant in France. Seeking papal approval for exhibition of the Shroud would have sent a signal to everyone avaricious enough to want it for themselves.

However, things had changed by the middle of the fourteenth century. The Hundred Years' War had weakened the authority of the French Kings. Philip VI and his son, John II, certainly did not have the strength of character or ruthlessness of their predecessors. The Black Death had ravaged France. With Geoffrey de Charny circumstances arose that permitted the public exhibition of the Shroud.

Geoffrey de Charny may have come into possession of the Shroud about 1350, when he married Jeanne de Vergy. At this time he was one of France's leading soldiers. He was a member of the King's Council and carried out delicate political negotiations on behalf of John II. He was even distantly related by marriage to the King. He was not a man to be trifled with. As we have seen, with the situation in France in the early 1350s, de Charny probably felt that exhibiting the Shroud was a positive and prayerful religious act which he could carry out without fear of being interfered with. There is no evidence that he obtained specific papal approval for this, but in 1354 he did petition Pope Innocent VI in Avignon for collegiate status of his church in Lirey. That same year Innocent VI granted indulgences to pilgrims visiting the church in Lirey[23]. This may be an indication that he knew of, and approved, the presence of the Shroud in Lirey.

The younger Geoffrey de Charny, while a respected man in France, did not have the authority or status of his father. It is possibly because of this that Bishop d'Arcis in 1389 made his apparent bid, both in a request to King Charles VI and in his purported Memorandum to the Pope, to take ownership of the Shroud. It is also quite possible that the same insecurity of possession led Marguerite de Charny to finally give the Shroud to a royal house that could be trusted to keep it safe in perpetuity, rather than leave it to relatives who could not be relied upon to keep it safe. Margaret had no children of her own to leave it to.

Having said all that, there is one indication that the Shroud was, somehow, known in France during the period before its appearance in Lirey. As we have seen, one theory regarding the transfer of the Shroud from Athens to France is that of Monsignor Arthur Barnes, who suggested that Otto de la Roche sent it to his family in France while he remained in Athens. Mgr Barnes went on to suggest that the family gave the Shroud to the Archbishop of the nearby city of Besancon, who put it in his cathedral where it remained for a century and a half, being exhibited every Easter and the focus of pilgrimages[24].

In a recent paper Prof Daniel Scavone has also referred to a possible identification of the Shroud with a cloth said to have been used in the Easter liturgy at the cathedral of Besancon in 1253. He noted that this hypothesis has been scrutinized by scholars but never refuted[25]. The public display of the Shroud in Besancon during the thirteenth century could also explain from where the Templar painter of the Templecombe Head drew his inspiration.

In 1349 the cathedral of Besancon was struck by lightning and burned to the ground. After this there was no trace of the Shroud that had reputedly been kept there. It was Mgr Barnes's theory that the Shroud was removed from the cathedral in the chaos by a member of the de Vergy family and taken to Calais where they gave it to King Philip VI. The King, in turn, entrusted the Shroud to Geoffrey de Charny, whom he expected to return it to its rightful owner in Besancon. De Charny in turn decided to keep the genuine Shroud and sent a painted copy to Besancon[26].

The Missing Years: From Athens to France

This all seems a little far-fetched and unnecessarily complex. From what we know of Geoffrey de Charny, he was a most unlikely person to have unlawfully kept a valuable artifact that had been entrusted to him by his king with an express purpose. If he had been given the Shroud in trust, there can be no doubt that he would have fulfilled the terms of the trust. Further, there is no evidence whatsoever linking the de Vergy family with any theft of a shroud from Besancon.

It is also unlikely that any King of France who came into possession of such an important relic as the very shroud of Jesus Christ would give it to one of his subjects, however exalted that subject might be. We have already seen that Louis IX had received a collection of relics, which included the Crown of Thorns, from Constantinople and placed them in the Sainte-Chapelle in Paris. If Philip VI had obtained the Shroud, it must be a certainty that he would have have kept it in Paris with other relics of this nature.

However, there was a painted Shroud in Besancon from at least the fourteenth century onwards, the origin of which is not known. It may have been a copy of the original Shroud dating from as early as the eighth century. It may also have been a copy of the Shroud at Lirey made in the 1370s[27].

A shroud showing the dorsal image only and eight feet in length is reported to have been venerated in Jerusalem at the start of the eighth century. In about 705 it was reported that a Bishop Arculf had been shown such a shroud in Jerusalem, which he had been told was the shroud of the sepulchre. The same story of a shroud in Jerusalem was told in the eleventh and twelfth centuries by a French monk called Bernard and then by Peter, the Deacon of Monte Cassino[28]. It is possible that this shroud was yet another of the "not made with hands" copies of the Cloth of Edessa that had been made in the sixth and seventh centuries.

Whether the shroud in Jerusalem ever found its way to Besancon is not known. However the Shroud of Besancon was also eight feet long and also showed the dorsal image only. It was described as being of "ordinary soft linen such as that made in Egypt"[29]. This shroud was destroyed at the time of the French Revolution. A copy of it is still retained in the Citadelle de Besancon[30]. According to local legend there were originally two shrouds, the Shroud of Turin, in which Christ was wrapped after being taken down from the Cross, and the Shroud of Besancon, in which he was wrapped on the way to the tomb. This legend does not explain how it was possible to wrap a man in a shroud measuring only eight feet long by four feet wide.

Could the reports of a shroud in Besancon in some way reflect knowledge of the existence of the Shroud of Turin in France prior to 1355, particularly as the home of the de la Roche family was close to Besancon? Again, this can only be a matter of speculation. Copies of the Shroud of Turin were made in Europe, such as the 1516 shroud in the Church of St Gommaire in Lierre in Belgium and a copy made around 1550 that is kept at Xenobregas near Lisbon in Portugal. If the Shroud of Besancon was not originally a Byzantine or Edessan copy of the Cloth of Edessa, then it may well have been a copy of the Shroud of Turin originating in France prior to 1355.

An intriguing addition to this question is the fact that there is also another copy of the Shroud of Besancon in Ray-sur-Saone Castle. This is in the form of a painting on canvas. There are some differences between the figure in this painting and the image on the Shroud of Turin. Both show the bruised cheekbones, the marks of the crown of thorns and the wounds on the left side. However the hands on the painting are crossed left on right and the feet are separated rather than overlapping. The hand wounds are in the palms rather than the wrists. The painter seems to have coupled observation of the Shroud with beliefs of the time. In this regard it is similar to other known copies of the Shroud of Besancon[31].

There does seem to be some link through Ray-sur-Saone between the Shroud of Besancon and the Shroud of Turin.

The final question is, when and why did the Shroud pass from the de la Roche/de Ray family to their de Vergy descendants rather than stay in the main line of de Ray descent (assuming that this is what happened)? The de Rays continued to be an influential family in Burgundy well into the fourteenth century. The de Vergys were also influential, with Jean I de Vergy becoming Seneschal of Burgundy in 1291, as mentioned above. If the Shroud was indeed passed down by Otto de la Roche to his descendants, one can only guess at how it passed to the de Vergys and then out of the direct de Vergy male line of descent to Jeanne de Vergy.

The Missing Years: From Athens to France

PART 3 –
THE SHROUD AND THE CRUCIFIXION

Chapter 10 – Death on the Cross

The image on the Shroud is clearly that of a crucified man. Further, it is the image of a man who was tortured and crucified in a manner very similar to that described in the Gospels with reference to the Passion and death of Jesus Christ. Whether the image is indeed that of the crucified Christ comes down to being a matter of individual belief. How the image was formed on the cloth will be addressed in detail in Part 4 of this book.

However, in order to better understand the Shroud and also to grasp what it may represent in terms of the Passion of Christ, it is necessary to have a clear understanding of crucifixion as a form of capital punishment at the time of the Roman Empire.

The word "crucifixion" comes from the Latin word *cruciare*, meaning to torture or torment, from which the English word "excruciating", used to describe severe pain, is also clearly derived. It was a hideous form of capital punishment used in ancient times by many civilizations in the Middle East and the Mediterranean Basin. These included the Persians, Phoenicians, Assyrians, Babylonians, Egyptians, Greeks, Seleucids, Carthaginians and Jews, as well as the Romans. It may have originated in Asia Minor about the sixth century BC with the Assyrians, Phoenicians and Persians. There were a number of well-recorded occasions when mass crucifixions took place. After the slave revolt led by Spartacus against Rome from 73 – 71 BC, six thousand of his followers were crucified along the Appian Way outside Rome, a scene memorably depicted in the 1960s film *Spartacus*. According to the Jewish historian Flavius Josephus, the Roman general Titus crucified five hundred Jewish prisoners each day at the time of his siege of Jerusalem during the Jewish War of 66 – 70 AD.

Crucifixion was eventually abolished in the Roman Empire by the Emperor Constantine the Great in 314. It continued however in the Middle East and in the Moslem world and has been used as a method of martyrdom of Christians, in imitation of the manner in which Christ was crucified. More recently, Jews were reported to have been crucified during World War II in Dachau concentration camp and there have been contemporary reports of crucifixions in Sudan[1].

Crucifixion was always regarded with great horror. Josephus reported that, during the Jewish War, just the sight of the Roman army preparing to crucify a popular Jewish captive would be sufficient to bring about the surrender of a citadel[2]. The Romans considered it a deterrent to crime and a punishment reserved primarily for criminals and slaves. It was Cicero's view that no Roman citizen should suffer the penalty of crucifixion, although there were numerous recorded incidents where citizens were crucified as punishment for high treason or crimes against the state[3]. The abhorrence of crucifixion was such that, although there are numerous references to it by Roman writers, no writer of the time ever provided any clear and full description of its physical details[4]. Even the Gospel writers gloss over the detail of Christ's crucifixion, leaving everything to the imagination of the reader. Luke, the most descriptive of the Gospel writers, for example, merely records that "when they reached the place called the Skull, they crucified him there and the two criminals also...."[5] The only reference to nailing in the Gospels is in John's Gospel where Thomas is recorded as saying, "Unless I see the holes that the nails made in his hands and can put my finger into the holes they made.... I refuse to believe.[6]"

Despite the tens or even hundreds of thousands of crucifixions that must have taken place over the centuries in antiquity, physical evidence of only one has ever been found. In 1968 during construction work at Giv'at ha-Mivtar, to the north of Jerusalem, a burial ground was found dating to the first century AD. In one ossuary there, bearing the name Jehohanan, the remains of a young man in his twenties were found, including a heel bone with an iron nail through it. He had clearly been crucified.

There are several reasons why more evidence of crucifixions has not been found. According to the Israeli scholar Joseph Zias, nails used in crucifixions were in great demand, because they were believed to have magical powers in warding off evil, in curing various medical conditions and as good luck charms. They were removed from their victims immediately after crucifixion or, alternatively, were stolen from the tombs of victims of crucifixion. Without the nails, there is no real physical evidence of crucifixion in human skeletal remains. During mass crucifixions victims were taken down after they had died and thrown in heaps, where their remains were devoured by scavenging animals and birds of prey. Nails were also retained by the executioners for subsequent crucifixions[7].

Roman crucifixions were usually carried out by teams of five soldiers. The team consisted of the *exactor mortis*, usually a centurion, who was in charge of four soldiers, called the *quaternio*. The teams were well trained and experienced in the techniques of crucifixion. Nailing of both the hands and feet is believed to have been the general rule, although in mass crucifixions, due to the time involved in nailing individuals to crosses, ropes were more commonly used. The *exactor mortis* also had the duty of determining if and when the victim was dead[8].

The most common form of cross used by the Romans was the *crux commissa*, or the T or *tau* cross. This consisted of an upright called the *stipes,* which was a permanent feature of the execution site. The crossbeam, known as the *patibulum*, was fixed to the top of the *stipes*, either by fitting it into a groove or by nailing or binding the two together[9].

In Jerusalem the execution site was set up just outside the city walls, at a place called *Calvarii* in Latin or *Gulgutha* in Aramaic, both of which mean "skull", probably because of the shape of the hilly knoll used for crucifixions. The place of crucifixion was close to the tombs, which facilitated transport and burial. The *stipes* on Calvary remained in place for a number of years after the crucifixion of Jesus, which probably took place in 30 AD. In 41 – 42 AD the course of the city wall of Jerusalem was changed as King Herod Agrippa I extended the perimeter of Jerusalem. Calvary was no longer outside the city wall, but just inside, so it could no longer be used as an execution site. It must be assumed that the *stipes,* or more likely the group of *stipes* on Calvary, was taken down at this time[10].

The Roman nail used in crucifixions was made of iron, with a square shaft gradually tapering from head to point. There is a relic in the Basilica of S. Croce in Gerusalemme in Rome that is said to be one of the nails used in the crucifixion of Jesus. It is 12,5 cm long with a square iron shaft measuring 9 mm across at the head and tapering gradually to the point area, where it measures 5 mm across. It bears a strong resemblance to the nail found with the remains of the crucifixion victim Jehohanan.

At the place of crucifixion the victim was stripped, if he was not already naked[11]. He was forced to lie on the ground with arms stretched along the *patibulum*. Two of the executioners held him down while a third, and possibly the fourth, nailed his hands to the piece of wood The two executioners then lifted the *patibulum* while a third raised the victim, or *crucarius*, to his feet by grasping him about the waist. The victim was then backed up to the *stipes*, where the *patibulum*, with the victim attached, was lifted into position over the *stipes*. Dr Pierre Barbet, writing in *A Doctor at Calvary*, suggests that the victim was forced backwards up a ladder leaning against the *stipes*[12], while two members of the *quaternio* lifted the *patibulum* and the third helped the victim up to the top step. Once the *patibulum* was in place the steps were pulled away while the victim's legs were held. His feet were then nailed to the *stipes*[13].

In view of the effort required to lift the *crucarius* on to the cross, it can be assumed that the cross was not high. The Roman cross probably stood no more than seven feet high. Some crosses contained a *sedile,* or saddle, attached to the *stipes*, which would allow the victim to support his body by sitting on it, to both ease and prolong his suffering[14]. Many mediaeval and modern paintings and crucifixes also show a support placed under the feet of Jesus, the *suppedaneum*, but there is no mention of this in early literature, early crucifixes do not show it and it appears to be an artistic invention[15].

* * * * * * * * *

The exact manner in which the *crucarius* was nailed to the cross has been the subject of recent speculation. Depictions of the crucifixion of Jesus have generally shown him as being nailed to the cross through the palms of the hands. Dr Pierre Barbet carried out experiments which, he claimed, proved that this would be impossible. He took an arm that he had amputated and drove a square nail into the middle of the palm, suspending the arm from a beam. He then suspended a weight of 88 lb from the elbow of the arm, this being about half the average weight of a man about six feet tall. After a short period of time the nail tore through the skin of the hand[16]. Dr Barbet took this as evidence that nails could not be driven through the palms of a *crucarius* without rapidly causing a tear. As a result of further experiments, he concluded that, for the *crucarius* to be secured to the cross, the nails had to be driven through a gap in the bones of the wrist, known as Destot's Space. Dr Barbet also noted that any nail passing through this gap would injure the trunk of the median nerve of the arm, causing extreme pain. He also found that one result of nailing in this fashion would be a reflex movement of the thumb into the palm of the hand[17].

Dr Barbet's work has recently been challenged by an American pathologist, Dr Frederick Zugibe. Dr Zugibe has shown that Dr Barbet made an error in his placement of Destot's Space and that in fact this gap does not coincide with the median nerve. Destot's Space is on the ulnar, or little finger, side of the wrist, not the radial, or thumb side[18]. Nailing through Destot's Space, therefore, would not affect the median nerve of the arm nor would it cause any reflex movement of the thumb.

Dr Zugibe has also shown that the position of the hand wound on the image on the Shroud of Turin does not correspond with Destot's Space and that Dr Barbet's hypothesis does not therefore coincide with the evidence of the Shroud.

Dr Zugibe's own hypothesis is that nailing would have been carried out through the upper part of the palm. He describes his hypothesis in the following terms:

"Touch your thumb to the tip of your little finger. A deep furrow called the *thenar furrow* is seen at the base of the bulky prominence extending from the base of the thumb. If a nail is driven into this furrow, a few centimeters from where the furrow begins at the wrist, with the point of the nail angled at about 10 to 15 degrees toward the wrist and slightly toward the thumb, the nail would be inclined toward the area created by the metacarpal bone of the index finger and the capitate and lesser multangular bones of the wrist, which I have coined the Z area.[19]"

He also shows that this corresponds with the hand wounds on the Shroud.

Dr Zugibe has also challenged Dr Barbet's conclusion that the palms of the hands will not support the weight of the body of a *crucarius*. Dr Barbet's experiment related only to a situation where the body would be hanging free. As the feet of the *crucarius* were also secured to the *stipes*, his experiment was invalid. Dr Zugibe carried out experiments in which he secured volunteers to a cross (naturally enough without nailing them to it!) and studied the forces involved. He reached the conclusion that the palms of the hand would support the *crucarius* when his feet are also secured[20].

Finally, Dr Zugibe noted that crucifixions were performed in many different ways. In the case of Jehohanan, the crucified Jew of the first century whose remains were found near Jerusalem, he appears to have been nailed between the radius and ulna bones in the forearm. Some nailing was done in this way, some through the palms, with or without rope supports, and some in unintended regions of the hands if the victim struggled[21].

The feet were secured to the *stipes* by bending the knees and sliding the soles up the *stipes* in order to lay the soles flat prior to nailing. As mentioned above, the idea of a foot support, or *suppadenaeum*, was the invention of later artists. The feet were nailed directly into the cross. Jehohanan appears to have been nailed through the side of his feet, with his feet astride the upright rather than in front of it. The question of whether Jesus was nailed with one nail through his feet or two has been the subject of religious controversy. There is a tradition dating back to Gregory of Tours in the sixth century that he was nailed with four nails, of which two were through the feet[22]. This tradition was ratified in the early thirteenth century by Pope Innocent III, when he condemned the beliefs of the Cathar heretics of southern France at the start of the Albigensian Crusade[23]. Their beliefs seem to have included a three-nail crucifixion[24]. At about the same time, however, the Landgrave Psalter, which is presently kept in the Wurttembergische Landesbibliothek in Stuttgart in Germany, also showed Jesus nailed by three nails, with one nail securing both feet[25]. This has now become the conventional way of depicting the crucified Christ.

The four-nail tradition lasted in Italy well into the fourteenth century, but disappeared elsewhere, despite the strictures of Pope Innocent III.

Dr Zugibe has noted that it does seem likely that, in most instances of crucifixion, each foot was nailed separately and flush to the cross. It would have been easier for the executioners to have nailed each foot separately rather than try to hold one foot on top of the other and, without solid support, drive a single nail through both feet, particularly if the victim was struggling[26]. The four-nail hypothesis may, therefore, be the correct one after all.

* * * * * * * * *

The process of crucifixion started with the victim being scourged. In Rome scourging was a legal preliminary to every execution, whether it was beheading or crucifixion. The only cases where a condemned person was exempt were senators, soldiers or women who had the freedom of the city. In the case of beheading the scourging was done with rods belonging to civil officers called *lectors*. In other cases scourging was carried out with a distinctive Roman instrument called the *flagrum*. This was a whip with a short handle and two or three, or even more, long, thick thongs. Towards the end of these thongs small balls of lead or the small bones of sheep were inserted. The thongs would cut the skin and the inserted balls or bones would rip the flesh, causing a great deal of bleeding.

While the Romans had no laws governing the number of lashes that could be given, Jewish law restricted the number of blows to 40. In fact the Pharisees, to be scrupulous, limited the number to 39.

Scourging was sometimes used by the Romans as a form of capital punishment, where the victim was beaten to death. In the case of crucifixion, however, some restraint was used so that the victim would not die too quickly on the cross[27].

The usual method of scourging was for the victim to be stripped naked and shackled by the wrists to a fixed object such as a low column. This caused him to be bent over. One or more soldiers carried out the punishment, lashing the flagrum across the naked back of the victim. The weight of the metal or bones in the thongs would carry them to the front of the body as well as down the sides and across the arms and legs as far down as the calves[28].

Following the scourging the condemned man went on foot and naked, carrying his *patibulum*, to the execution site, where the *stipes* awaited him.

In deference to Jewish customs, Jewish *crucarii*, such as Jesus, were spared the humiliation of walking naked, but were stripped when they arrived at the *stipes*[29]. For this last journey the *patibulum* was placed on the shoulders and outstretched arms of the victim, after which it was frequently bound with cords to his chest, arms and hands. Thus if the victim fell while carrying the beam, he would be unable to protect his face or body from the fall. The *patibulum* probably weighed in the region of 70 - 100 pounds, so the effects of a fall would be extremely painful and damaging.

Finally, the bearer of the cross was preceded by, or had hung around his neck, an inscription on wood giving his name and the crime for which he was condemned. This was known as the *titulus* and was later fastened on to the cross[30].

On arrival at the execution site the *crucarius* was nailed to the *patibulum* and then hoisted on to the *stipes* as described above. As has been mentioned, the *quaternio*, the execution team, would have been experienced and the actual process of nailing and hoisting would have been carried out quickly and efficiently. The victim would have been prevented from struggling. Once the *patibulum* was in place, the victim's feet would then be secured to the upright. As also mentioned above, if it was intended to prolong the agony of the victim on the cross a seat or saddle, known as the *sedile*, was attached to the upright, on which the victim could rest.

The nailing itself would have caused violent and extreme pain. Nailing through the *Z* area of the wrist or through the palm, as described by Dr Zugibe, would injure the median nerve, causing an extremely painful condition called *causalgia*. This pain has been described as "traversing the arm like lightning bolts"[31]. Victims frequently go into shock and the pain is exacerbated by increased temperature and direct exposure to the sun.

The nailing of the feet would have a similar effect due to damage to the plantar nerves. These nerves bifurcate down the foot with a branch on both sides of each metatarsal bone. Even the slightest injury to these nerves would cause *causalgia*[32].

Once on the cross, the victim was left to die in agony. The execution team would remain on guard by the cross to prevent the condemned man from being rescued by friends or sympathizers.

* * * * * * * * *

The cause of death on the cross has also been the subject of medical controversy. The most widely disseminated theory has been that of Dr Barbet, who suggested that crucifixion caused asphyxia. He referred to the work of an earlier French doctor, Dr Le Bec, who suggested that the raised position of the arms of the *crucarius* would cause a relative immobility of the sides of the body and would therefore hinder the ability to exhale. He also referred to a severe punishment in the Austro-German army in World War I, called *aufbinden*, in which the condemned man was hanged by two hands from a post with the tips of his feet barely touching the ground. This led to cramps and a situation where the lungs could not expel air. The normal oxygenation of the circulating blood does not take place and the victim starts to asphyxiate. Dr Barbet also cites a similar punishment, with similar effects, having been carried out in the World War II German concentration camp at Dachau. In the latter case the feet were some distance from the ground and the condemned were left hanging until they died[33].

From these observations Dr Barbet concluded that the crucified all died of asphyxia, after a long period of struggle. He further suggested that the *crucarius* sought relief from the effects of asphyxia by using his feet, which were fixed to the *stipes*, as a fulcrum to lift his body and relieve the pressure on his arms. However, as his legs became tired, he would again sag on the cross, bringing on a further attack of asphyxia. As a result the process on the cross became one of alternatively sagging and then straightening the body, with alternating periods of asphyxiation and respiration.

The executioners could, if and when they wished, end this agonizing cycle by breaking the legs of the *crucarius*, an action called the *crurifragium*. Apart from the extreme pain caused by this, it would also have the effect of preventing the *crucarius* from lifting himself to relieve the effects of asphyxiation. He would therefore die quite rapidly from asphyxiation following the *crurifragium*[34].

Dr Zugibe has pointed out that Dr Barbet has drawn his conclusions from observations of men whose arms were suspended directly above their heads and whose legs were hanging free. These conclusions would not be applicable to suspension on a cross. Dr Zugibe carried out tests with volunteers who were secured to a cross in the position of a Roman *crucarius*. He found no evidence of severe respiratory distress. All of the volunteers reported a severe pull on their shoulders; tightness, pain or cramping in the calves; or pain in the knees, feet and wrists. Their breathing tended to be abdominal, with short inspirations and longer expirations, a sort of gasping rather than inability to exhale.

Rather than raising themselves on their feet, the volunteers tended to seek relief from their discomfort by arching their backs. In fact none of them were able to straighten their bodies on the cross in the manner suggested by Dr Barbet.

Dr Zugibe observed that the oxygen content of the blood of the volunteers either remained constant or actually increased slightly, which is quite contrary to what would be expected if asphyxiation was taking place. Heart beat and blood pressure both increased as well[35].

Dr Zugibe is therefore firmly of the opinion that asphyxiation would not be the cause of death of a *crucarius*. It is his view that death would be caused by shock. There are various types of shock, of which two – hypovolemic shock (low volume shock) and traumatic shock (injury shock) – would specifically affect a person who underwent the total process of crucifixion[36].

As Dr Zugibe describes it, "The... scourging with a *flagrum* composed of leather tails containing metal weights or bone at the tips would have caused penetration of the skin with trauma to the nerves, muscles and skin; rib fractures or cartilege/rib dislocations, lacerations, infiltration of significant amounts of blood throughout the intercostal spaces and back and chest musculature, bruises and alveolar (air spaces in the lungs) rupture; and very probably a collapse of a lung (pneumothorax), reducing the victim to an exhausted, wretched condition.[37]"

The result would have been significant blood and fluid losses and severe pain, leading to both hypovolemic and traumatic shock at this early stage.

After the scourging the victim then had to carry, in his weakened state, a heavy beam of wood some distance on foot, over a rough path. The chances of stumbling or falling would have been high. With his arms tied to the wooden cross-beam, the *crucarius* would have been unable to break any fall and would have crashed into the ground, leading to yet more pain and trauma. Again, Dr Zugibe's description is graphic:

"....carrying the crosspiece on the shoulder for a short time, falling some of the time, and sustaining blows at other times also would have added to the hypervolemic and traumatic shock. Pleural effusion (water round the lungs) would gradually have increased following the brutal scourging and begun to compromise breathing with shortness of breath. The haemorrhages from rib fractures and lung lacerations would have added to the hypovolemia. Increasing hypovolemia would have resulted.[38]"

At the execution site, the act of driving nails through the victim's hands and feet would have caused the horrific pain of *causalgia*. This would cause additional traumatic shock and hypovolemia.

On the cross itself, the *crucarius* would have experienced episodes of agonizing pain from all of his injuries. These episodes of pain and the unrelenting pains in the chest wall from the scourging would greatly increase the level of traumatic and hypovolemic shock. In addition, in a Mediterranean climate the *crucarius* would be exposed to the heat of the sun. Both hypovolemic shock and the heat would result in severe thirst.

With no respite, the degree of shock incurred by the victim would eventually become fatal and the victim would die from cardiac and respiratory arrest brought on by hypovolemic and traumatic shock.

Dr Zugibe further suggests that the breaking of the victim's legs, the *crurifragium*, was carried out as a *coup de grace* when the victim was near death. This would cause severe traumatic, haemorrhagic shock as a result of both pain and blood loss. In some cases a fatty embolism could be caused by the breaking of the legs, resulting in rapid death. Dr Zugibe's own words once again illustrate the situation:

"Since the *crucarius* was in a severely weakened condition due to shock, the marked haemorrhage from the breaking of the legs and the severe pain would deepen the level of hypovolemic and traumatic shock, with a consequent drop in blood pressure and rapid development of congestion in the lower extremities, resulting in unconsciousness, coma and death.[39]"

The *crurifragium* probably also had a secondary function – to prevent the *crucarius* from crawling away from the cross if, by any chance, he was still alive if and when taken down. It has been suggested that the breaking of the legs also served to gratify the spectators.

Once the *crucarius* was dead, the usual course was for his corpse to be left on the cross to be devoured by birds of prey and wild beasts. However, the bodies could be asked for by families who wished to give them a proper burial. This was usually permitted without any hindrance or demand for payment[40]. Once the body had been removed, either for burial or by the forces of nature, the *stipes* awaited its next victim.

There is no information in ancient literature about the manner in which a *crucarius* was removed from the cross after death. It can be speculated that a rope would have been tied around the waist of the corpse and over the crosspiece, to hold the body in place while the nails were removed from the wood. This could have been done quite easily with a few slight taps on the points, which would presumably have been protruding through the wood. The *crucarius* could then be lowered to the ground and the nails removed from his hands and feet[41]. Rigor mortis would have set in quite quickly on the cross and the body would almost certainly, at the time of removal, be in a state of rigor. Rigor mortis is a stiffening and shortening of all the muscles of the body after death caused by an irreversible chemical reaction[42]. The stiffness of the body would therefore have been an additional difficulty in removing it from the cross for burial.

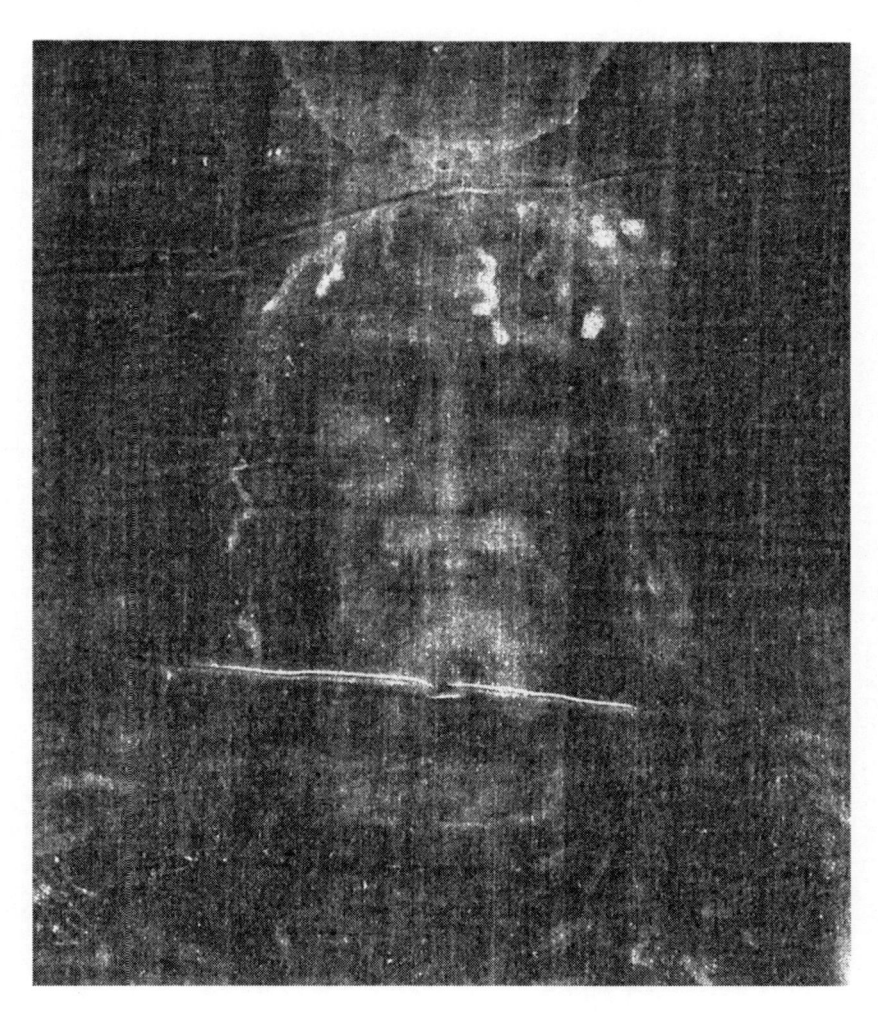

The Man on the Shroud – Negative Facial Image.
Copyright Barrie M Schwortz.   Permission granted.

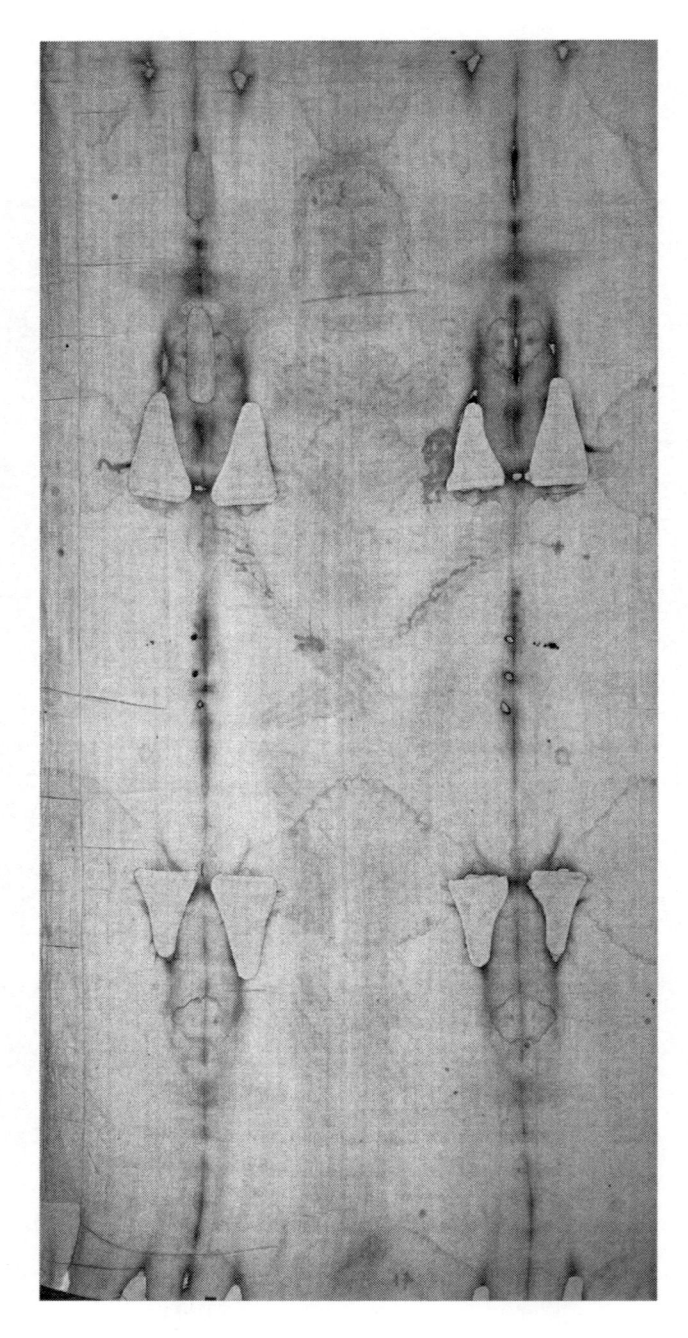

The Dorsal Image on the Shroud - Natural.
Copyright Barrie M Schwortz. Permission granted.

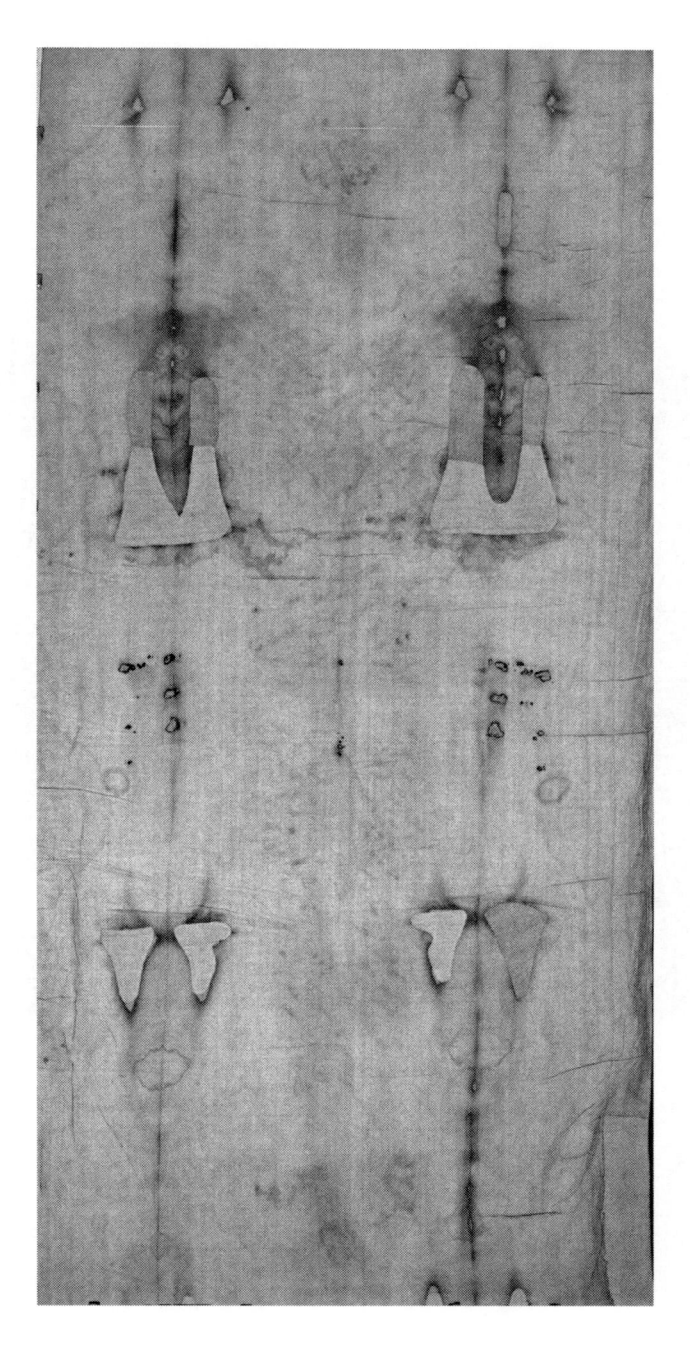

The Ventral Image on the Shroud - Natural.
Copyright Barrie M Schwortz. Permission granted.

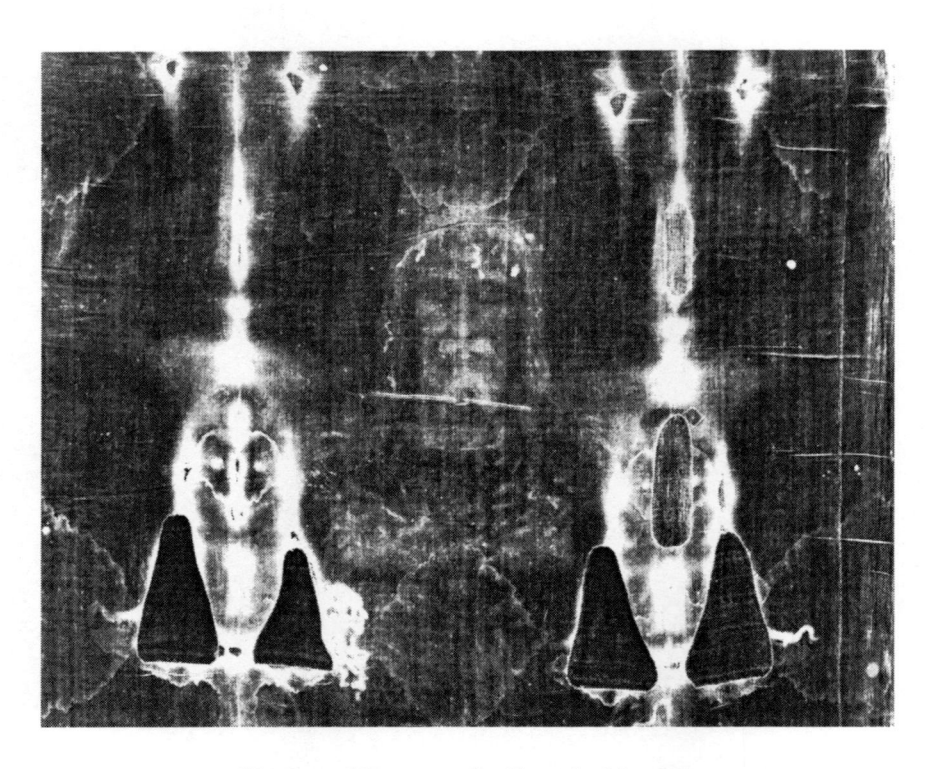

*The Dorsal Image on the Shroud – Negative.
Copyright Barrie M Schwortz. Permission granted.*

The Ventral Image on the Shroud – Negative.
Copyright Barrie M Schwortz.   Permission granted.

*Painting of the Shroud of Besancon at Chateau Ray-sur-Saone.
Copyright Alessandro Piana. Permission granted.*

The Chest at Chateau Ray-sur-Saone.
Copyright Cesar Barta. Permission granted.

*Chateau Ray-sur-Saone with Cesar Barta in the Foreground.
Copyright Cesar Barta. Permission granted.*

The Pilgrim's Medallion.
Public Domain.

Geoffrey de Charny.
Public Domain.

Anti-Pope Clement VII.
Public Domain.

*King John II of France.
Public Domain.*

The Porte d'Oriflamme.
Public Domain.

The Battle of Poitiers.
Public Domain.

Aerial View of Lirey Today.
Public Domain.

The Church at Lirey Today.
Public Domain.

The STURP Team's First Look at the Shroud – Turin 1978.
Copyright Barrie M Schwortz.   Permission granted.

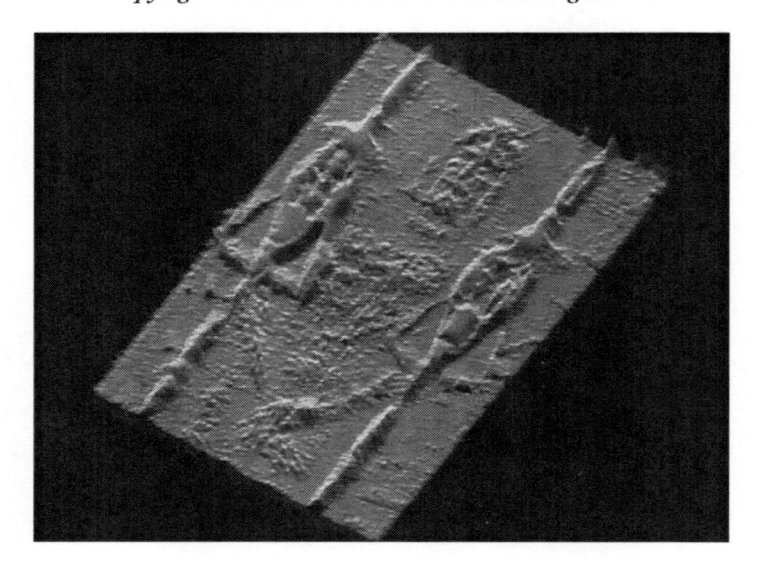

Shroud ¾ View on the VP8 Screen.
Copyright Barrie M Schwortz.   Permission granted.

*Mural of the Good Shepherd in the Catacomb of St Callixtus, Rome.
Authority received from the Pontifical Commission for Sacred Archaeology.*

*Mural of the Good Shepherd in the Catacomb of St Callixtus, Rome.
Authority received from the Pontifical Commission for Sacred Archaeology.*

*Mural of the Good Shepherd in Lincoln Cathedral, Lincoln, England.
Photograph by the author.*

*Bearded Christ in the Catacomb of Commodilla, Rome.
Authority received from the Pontifical Commission for Sacred Archaeology.*

*Christ Pantokrator in the Monastery of St Catherine, Sinai.
Public Domain.*

The Holy Face of Laon.
Public Domain.

The Templecombe Head.
Public Domain

Jesus – 1935 Painting by Agemian.
Copyright the Confraternity of the Precious Blood. Permission Granted.

*Christ Pantokrator in the Dome of the Greek Orthodox Cathedral,
Maputo, Mozambique. Painted by Helen Lieros.
Photograph supplied by the artist.*

Chapter 11 –
The Passion and Death of Jesus Christ

Jesus's teaching had greatly disturbed the Jewish religious leaders in Jerusalem. He had been highly critical of their slavish adherence to religious law and customs and had caused deep offence by mixing with those seen by the religious leadership as being beyond redemption, such as tax collectors and prostitutes. He had created disturbances in the Temple. At the same time his teachings had great appeal to the poor and less privileged of society. At a time when political radicals and hotheads were active, Jesus must have been seen by the religious establishment as another threat to order and stability. As Luke describes it in his Gospel, "the chief priests and scribes were looking for some way of doing away with him."[1]

Jesus's Passion, the period of intense suffering before his death on the cross, started at the Passover meal that he shared with his apostles on the evening before his death. St John's Gospel describes him as being troubled in spirit[2]. He knew that Judas Iscariot would betray him. The Last Supper was the last occasion that he would have to speak to his apostles and leave his teaching with them. As John describes in his Gospel, Jesus used this time for a lengthy discourse in which he summed up his teaching and took his final leave of his friends.

From the Last Supper he went to the Mount of Olives outside Jerusalem, where he went into a small estate called Gethsemane[3]. There he withdrew from the apostles and knelt down to pray. He was in great distress, to such an extent that "his sweat fell to the ground like great drops of blood."[4]

This corresponds to a recorded although rare medical condition known as *hematidrosis*, which has been defined as an excretion of blood or blood pigments in the sweat[5]. One of the earliest descriptions of this phenomenon was by Aristotle. Recent cases have been reported in the United States in the late nineteenth century and in France, Russia, Germany and Britain on various occasions in the nineteenth and twentieth centuries. Writers on the subject have concluded that many cases of *hematidrosis* have been associated with a severe anxiety reaction triggered by fear[6]. In particular, cases have been recorded among condemned prisoners being taken to execution.

Jesus was afraid. He knew what faced him. He knew what agony he was about to suffer. And he was afraid. He was overwhelmed by extreme fear, anxiety and sadness. The fact that he was on his knees is significant. It was unusual for a Jew to kneel down to pray. His weakness and his fear caused him to fall to his knees. His heart would have pounded against his chest. He would have broken out in a cold sweat. His muscles would have tightened and he would have trembled. This period of fear, sadness and distress clearly lasted some time – probably two hours or more. Three times he prayed and then went back to his apostles, each time to find them sleeping[7].

Jesus got no sleep that night. The period of prayer and fear was brought to an end by the arrival of Judas and the Temple guards, who arrested him and took him to the house of the high priest. Even at this tense moment and despite his weakness Jesus was able to carry out an act of healing, on the ear of the servant of the high priest[8]. Luke's account of Jesus's arrest emphasizes that Jesus maintained control over what was taking place, restraining his followers and talking calmly to those who had come to arrest him.

The olive grove of Gethsemane was east of the Cedron brook, a short distance up the Mount of Olives. It was only about 200 metres from the Temple enclosure and probably no more than half a kilometre from where Jesus and his apostles had celebrated their Last Supper[9]. The journey to the house of the high priest Caiaphas was a longer and more tortuous one. Jesus and the party who had seized him would have descended the slope of the Mount of Olives and turned south along the bed of the Cedron. The ground was undoubtedly rocky and uneven. Manhandled and pushed by his captors, Jesus must have stumbled from time to time and sustained grazes and bruises. After about 15 – 20 minutes the party would have turned and, after a steep ascent, entered Jerusalem through the Fountain Gate, to reach the house of Caiaphas. The distance traveled from Gethsemane was over a kilometre[10].

By the time that Jesus arrived at the house of the high priest it was the early hours of the morning. Already bruised and tired, he seems to have been immediately subjected to interrogation by the high priest and the Sanhedrin. Various witnesses were brought before him but with little effect. Finally the high priest asked him the straightforward question, "Are you the Christ, the Son of the Blessed One?" Jesus replied that he was, an admission that was seen by the priests and elders as blasphemous. They were unanimous that he should die[11].

This interrogation by itself must have been extremely stressful. Jesus was alone, a captive among people who were openly and aggressively hostile to him. Clearly his life was at stake. He had no opportunity to rest or even to try to gather his thoughts. Fear and loneliness must have been the foremost emotions in his mind.

During the hours that followed he was mocked and beaten by his guards. They blindfolded him, hit him and then asked him to play the prophet and tell them who had hit him[12]. Until daybreak he was tormented, beaten and insulted, a lonely, frightened man who faced a day of humiliation, abandonment and unspeakable agony.

Once day had broken, Jesus was taken on another journey – approximately a kilometre to the Antonia Palace – to appear before the Roman procurator, Pontius Pilate.

* * * * * * * * *

Pontius Pilate had become procurator of Judaea in 26 AD. He is a controversial figure. Jewish writers, such as Philo of Alexandria and Josephus, are extremely critical of him. Josephus' portrayal of Pilate has been described as "little short of character assassination"[13]. A typical example of the Jewish attitude to Pilate is given in a letter to the Emperor Caligula by the Jewish King Herod Agrippa I, who described Pilate as "of an unbending and ruthlessly hard character". He encouraged "corruption, violence, robbery, oppression, humiliations, constant executions without trial, and unlimited intolerable cruelty."[14]

Pilate was undoubtedly a man of his time. He must have been of equestrian rank, in other words among the Roman aristocracy, to have obtained the position of procurator. He held office for an unusually long time – ten years – and, unusually for a procurator, he had supreme judicial power. He is thought to have owed his position to the imperial counsellor, Sejanus[15], which would indicate that he had at least a degree of political influence at the imperial court of Tiberius in Rome. His loyalty was to his Emperor, and his actions would have been dictated by his loyalty. Minor local agitators would have had little significance for him, except for how they might affect Roman rule in the province of Judaea.

Pilate's troubles in Judaea started shortly after his arrival in the province. Soldiers had brought statues of the Emperor into Jerusalem as a result of which a large delegation from the city went to Pilate at his headquarters in Caesarea, to protest and seek the removal of what were seen as heathen images. According to Josephus Pilate threatened to kill them all. The Jews then flung themselves on the ground and exclaimed that they were ready to die rather then to transgress their Law. Impressed by what he saw as intense religious zeal Pilate then ordered the immediate removal of the standards from Jerusalem[16].

On a later occasion he spent sacred treasury funds on the construction of an aqueduct to bring water to Jerusalem from a distance of seventy kilometers. From a Roman point of view this probably seemed a wise and benevolent act on the part of the Roman administration. Somehow, however, it offended the Jews, resulting in riots and the deaths of a number of people[17].

Pilate therefore had every reason to believe that the Jews were difficult people to deal with. His responsibility to the Emperor in Rome was to maintain peace and ensure that Judaea continued to contribute to the imperial treasury. Civil disorder was to be avoided and Pilate knew from personal experience that Jewish religious sensibilities in particular made civil disorder a constant threat.

On a Friday morning some four years into his tenure of office, he was confronted by the Jewish religious leadership demanding that he condemn a religious teacher who had been causing controversy among the Jewish people. The real complaint of the Jewish leadership against Jesus was the question of blasphemy – a purely religious issue. However, in order to obtain Pilate's sympathy towards their objective, they accused Jesus of inciting revolt, opposing payment of tribute to Caesar and claiming to be a king – purely political offences[18].

Pilate was not fooled. He clearly realized that Jesus was innocent of the charges being laid against him but he could not simply dismiss the Jewish leadership and their allegations. His first move was to attempt to pass responsibility to Herod Antipas, the tetrarch of Galilee, who was in Jerusalem at the time, on the grounds that Jesus was a Galilean and hence came under Herod's jurisdiction[19]. Herod simply mocked Jesus and sent him back to Pilate.

Pilate's next step was to try to make use of a custom of the Jews whereby a prisoner was released on the Passover in recollection of the release of the Israelites from captivity in Egypt[20]. He offered to release Jesus on this basis but the Jewish leadership again thwarted him by calling for the release of a well-known brigand[21], referred to in the Gospels as Barabbas.

Pilate made it clear to the Jewish leadership that he could find no fault with Jesus and he dismissed their allegations in a manner bordering on contempt[22]. However, political considerations meant that he could not just release Jesus. He must have had some concern that to totally ignore the feelings of the Jewish leadership would cause a riot and civil disorder, in the same manner as had the incidents of the imperial standards in the city and the use of sacred funds for an aqueduct. There may have been other, similar incidents during Pilate's tenure of office which have gone unrecorded. As a sop to their feelings therefore, Pilate ordered Jesus to be scourged[23].

At this stage Jesus had had no sleep during the previous night. He was weary from having been dragged from Gethsemane to the house of Caiaphas, from there to the Antonia and from the Antonia to Herod's palace and back. He had been pushed and manhandled over a distance of several kilometres; he had been beaten and tormented by the retainers at Caiaphas's house; he had been mocked and insulted by Herod; and he had twice been brought before the supreme Roman authority in the province. Physically and mentally he must have been exhausted and drained; yet the agony was only now about to begin.

* * * * * * * * *

In accordance with Roman practice in carrying out a scourging Jesus would have been stripped naked before being shackled – a further humiliation for a devout Jew. The scourging would have been carried out at the Antonia, so he did not have to be taken far.

One or two Roman soldiers would have been given the task of carrying out the scourging. The thongs of the whip would have had pieces of bone or metal balls embedded at the end of each tail. As these weighted thongs smashed into Jesus's flesh he would have twisted in agony and fallen to his knees. The blows would have caused severe lacerations in Jesus's flesh as well as rib fractures. He would have suffered excruciating rib pain every time he tried to take a breath. There would also very likely have been damage to the lungs, causing even more pain and difficulty in breathing. Partial or complete collapse of a lung (pneumothorax) would have been a possibility.

One effect of a rib fracture is internal bleeding. Jesus would have lost copious amounts of blood both from the external bleeding caused by the beating as well as internal bleeding caused by internal injuries. During the scourging he would have vomited and suffered tremors and seizures. He would also have sweated greatly. By the end of the scourging Jesus would have been in an early stage of shock. Subsequent to the scourging fluid would have started to accumulate around his lungs, adding yet further to his breathing difficulties. There may also have been damage to internal organs such as the liver and spleen[24].

Jesus's scourging was almost certainly particularly severe. Pilate had not ordered it as a specific prelude to crucifixion but rather as a punishment in itself. While his objective would not have been to beat Jesus to death, he would have intended that the punishment would have been severe enough to appease the Jewish leadership.

Medically, Jesus's condition after the scourging would have been described as serious. The injuries caused by the scourging would have resulted in traumatic shock, and hypovolemic (low blood volume) shock would have started to develop as a result of blood loss. As a result Jesus would have felt weak and light-headed; his skin colour would have turned ashen and he would have had bouts of profuse sweating. The trauma to his chest would have caused an accumulation of blood, fluid and mucus, a condition described as "traumatic wet lung"[25].

Following the scourging, according to John, "...the soldiers twisted some thorns into a crown and put it on his head, and dressed him in a purple robe. They kept coming up to him and saying, 'Hail, king of the Jews!'; and they slapped him in the face."[26]

Following the pain of the scourging Jesus now had a new and equally horrifying pain to bear. The crowning with thorns would have caused a condition caused *trigeminal neuralgia*.

The *trigeminal* nerve supplies pain perception to the front half of the head. The fine branches of the nerve spread throughout the scalp. Irritation of this nerve causes bouts of stabbing, shooting and explosive pain to the right and left half of the face. The penetration of the thorns into Jesus's scalp would have had precisely this effect. Trigger zones for pain attacks are usually on the lip or the side of the nose. The effect on Jesus, therefore, of being slapped in the face after the crown of thorns had been placed on his head would have been paroxyms of severe pain. These could well have been acute enough to have immobilized him. Patients suffering from *trigeminal neuralgia* have described their pains as "knife-like stabs", "electric shocks" or "jabs with a red-hot poker"[27]. One medical specialist has described the pain in the following terms:

"*Trigeminal neuralgia* is said to be the worst pain that man is heir to. It is a devastating pain that is just unbearable in its several forms."[28]

The paroxyms of acute pain from the crown of thorns would have continued throughout the journey to Calvary and during the crucifixion, activated by the actions of walking and falling; from the pressure of the thorns on the cross; and from the continuous pushing and blows of the soldiers[29]. The effect of the crown of thorns would have been to deepen the level of traumatic and hypovolemic shock that Jesus was suffering.

By the time that Jesus returned to Pilate he must have been a pitiable sight. Beaten, his body distorted with pain, his vision blurred, covered with blood, sweat and vomit, he would have been barely able to stand. Pilate led Jesus out again before the Jewish religious leaders, possibly hoping that the sight of him would be sufficient to stir compassion and mercy. He was wrong; as John describes, "... when they saw him the chief priests and the guards shouted, 'Crucify him! Crucify him!'"[30]

Pilate obviously felt that he was left with no alternative. To have released Jesus in the face of Jewish demands for his death would have been to risk civil disorder and upheaval. He probably felt that he had done his best for the Galilean teacher. The fact that he was condemning an obviously innocent man would, to Pilate, have meant little in the greater scheme of things – he had his responsibility to the Empire, and that took preference over all other considerations. Accordingly, Pilate acceded to the Jewish demands and ordered Jesus's crucifixion. He had the last word, however. On the *titulus* he had written, "Jesus the Nazarene, King of the Jews", effectively mocking those who had called for Jesus's death. Despite protestations he would not have the wording on the *titulus* changed[31].

* * * * * * * * *

By this time it must have been approaching mid-morning. Jesus set out on his final journey as the day was starting to heat up, carrying a wooden beam, the *patibulum*, weighing about 70 – 100 pounds. He would not have been kept waiting long once the order for his crucifixion had been given. The *quaternio* would have been selected and given its orders with the minimum of delay. Roman military discipline would have ensured that there would have been rapid arrangements made and the condemned sent quickly on his way to the execution site.

The journey from the Antonia to Calvary started along one of the main streets of Jerusalem and the total distance to be covered was slightly less than a kilometre. To Jesus, in his condition, it must have seemed never-ending. The road was unpaved and uneven, with ruts made by carts. At the edge of the city the road started to climb. It is hardly surprising that Jesus had to be assisted in carrying the *patibulum*. Luke describes how, as Jesus was being led away from the Antonia, in other words right at the start of his journey to Calvary, a man coming in from the country, Simon of Cyrene, was made to shoulder the cross and carry it behind Jesus[32]. This reference of Luke's is totally consistent with the weakened physical condition Jesus must have been in at the time. He would not have been physically capable of carrying a 70 lb wooden beam over a distance of nearly a kilometre. It was the responsibility of the *quaternio* to ensure that the condemned man reached the execution site alive. In the case of Jesus they may well have seen that to make him carry the *patibulum* himself would have been fatal.

There is no reference in the Gospels to Jesus falling on the way to Calvary, although it is a part of Christian tradition. However, it would have been remarkable if he had not stumbled and fallen, even without carrying a heavy wooden beam, on the uneven, climbing path to the execution site.

Following the scourging and his being dressed in a purple robe for the purpose of mockery, Jesus had had his clothes restored to him. He did not make the journey to Calvary naked. This, however, must have led to another source of extreme pain. On arrival at Calvary Jesus was stripped of his clothing. His cloak or tunic would have stuck to the blood covering his body. It must have been torn off his body prior to crucifixion, like a giant plaster strip attached to his whole body.

By the time he arrived at Calvary, Jesus would have been almost numb with exhaustion. His breathing would have been short and excruciating; he would have been sweating in the heat of the noonday sun; the crown of thorns would have been sending jolting pains through his head and face; he would have been bruised and grazed from numerous falls; his whole body must have been a mass of intense pain.

At Calvary Jesus would have been thrown to the ground and made to lie on his back, with his arms stretched across the *patibulum*. One of the *quaternio* would have lain across his chest to immobilize him and another would have held his legs down, so that a third could nail his hands to the beam. With his chest already severely injured, this in itself would have been a source of agony and extreme breathing difficulties. Then large, square spikes were driven through the lower palms of each hand. As the median nerves in each arm were damaged by the nailing, the pain would have been brutal. Jesus would have screamed in agony, and probably not for the first time.

Once he was secured to the *patibulum* he would have been lifted and hoisted on to the *stipes*. The torture of the nailing would have been repeated on his feet, with equally horrifying pain from the damage to the plantar nerves.

At this stage Jesus would have been suffering from exhaustion, severe chest trauma, blood loss, *trigeminal neuralgia* and *causalgia* from the nailing of his hands and feet. His degree of shock would have been increasing and his heartbeat would have been rapid. He would have been light-headed and short of breath. The agony would have been increased by the heat of the sun, the direct sunlight and the pressure of the nails rubbing against the nerves in his hands and feet. It is perhaps worth noting that, far from pain in one part of the body relieving pain elsewhere, people suffering multiple pains experience a magnification effect rather than an additive effect[33].

* * * * * * * * *

On the cross there was further emotional suffering for Jesus, as well as the extreme physical pain that he was experiencing. His mother was present to see the suffering of her son[34], together with other women who had accompanied Jesus from Galilee and at least some of his disciples. That alone must have caused him considerable mental anguish. In addition he was mocked and taunted by both onlookers and the soldiers. Even one of the criminals crucified with him took the opportunity to abuse him[35].

According to Mark, in his Gospel, it was the third hour when Jesus was crucified – probably some time after nine in the morning. He died at the ninth hour – some six hours later[36]. Those six hours must have been an eternity of shattering pain and agony. The weight of Jesus's body would have pulled on the nails in his hands and feet, causing excruciating episodes of *causalgia*. There would have been unrelenting pain in his chest wall both from the scourging and from increasing pleural effusion and pulmonary edema. Jesus would undoubtedly have arched his body from time to time in an attempt to straighten his legs and relieve the cramps in his calves, arms and shoulders. This would have pressed his head, with the crown of thorns, against the *stipes*, reactivating bouts of *trigeminal neuralgia*. All of this would have greatly worsened his state of traumatic and hypovolemic shock[37].

One result of the shock, increased by the heat of the day, would have been extreme thirst. John notes how Jesus complained of thirst and was offered a vinegar-soaked sponge on a stick[38]. This was not, as many believe, an additional torment, but a genuine attempt to relieve his thirst. The vinegar offered to him would have been the sour drink commonly consumed by Roman soldiers. Undoubtedly it would have been there to relieve the thirst they would have felt in the hot sun.

After an extended period of continuous torment and suffering from late evening on the Thursday until at least mid-afternoon on the Friday, Jesus died. According to Dr Frederick Zugibe, the cause of death would have been:

"Cardiac and respiratory arrest, due to hypovolemic and traumatic shock, due to crucifixion."[39]

As the afternoon drew on, it became necessary for those on the crosses to be taken down, as it would not have been acceptable to the Jews to have had them remaining there during the Sabbath. The soldiers were instructed to deliver the *crurifragium*, but they found that Jesus was already dead. As John describes it:

"When they came to Jesus, they found he was already dead, and so instead of breaking his legs one of the soldiers pierced his side with a lance; and immediately there came out blood and water."[40]

This is quite feasible from a medical point of view. There would have been a steady accumulation of fluid in Jesus's chest around his lungs (pleural effusion) following the scourging. To produce the flow of blood and water the spear thrust would have had to have penetrated the pericardium, the sac containing the heart and the roots of the great blood vessels, and the right atrium, the right upper chamber of the heart. This would have been filled with blood because, just prior to cardiac arrest, the heart contracts and ejects the blood into circulation for the last time. As the spear was jerked out of Jesus's chest, blood would have flowed from the right atrium and fluid would have been released from the pleural cavity at the same time, resulting in the phenomenon of "blood and water".[41]

After violent death rigor mortis sets in rapidly. This is a stiffening and shortening of all of the muscles of the body after death caused by a chemical reaction. In Jesus's case, his body was in the hot sun and he had died a particularly violent and cruel death. Rigor mortis would have set in, starting in the jaws and neck and moving down to the shoulders and arms, possibly within an hour. In addition, the violence of his death would have caused the muscles of his neck to contract solidly, through what is known as a cadaveric spasm[42].

Jesus's body had to be removed from the cross quickly. One of his disciples, Joseph of Arimathea, asked Pilate to let him remove the body and he had it buried in a new tomb close to the execution site[43]. Removing the body in a state of rigor would not have been easy. Probably the nails would have been removed first from Jesus's feet and the *patibulum* with the body still nailed to it through the hands lifted down from the *stipes*. The nails through the hands would then have been removed. Assuming that rigor had already spread to the shoulders, the arms would have become rigid in their outstretched position. In order to carry the body to the nearby tomb the arms would have had to be forcibly bent at the shoulders, a process known as breaking the rigor[44].

In his Gospel John describes how Jesus's body was wrapped with about a hundred pounds of myrrh and aloes, and he emphasizes that what was done was done according to Jewish burial customs[45].

Chapter 12 –
The Evidence of the Man on the Shroud

The Shroud of Turin has been described by many writers. It is a linen cloth, ivory-coloured with age, with a three-to-one herring-bone weave, bearing the faint imprint of the back and front of a man. The man appears to be tall and strongly built, and he has long hair and a beard. He is naked and laid out in the attitude of death[1]. Following its restoration in 2002, it has been measured as follows: The bottom length is 14,4848 feet, or 441,5 centimetres, the top length is 14,2552 feet, or 434,5 centimetres, the left side is 3,7073 feet, or 113,0 centimetres, and the right side is 3,7303 feet, or 113,7 centimetres[2]. To put it more simply, it is a rectangular sheet measuring approximately 14 feet 3 inches long by 3 feet 7 inches wide. However, it is not exactly symmetrical and its measurements can vary depending on the tensions on the cloth and how it is stretched to be measured. The man was tall; his height has been estimated at 72 inches, plus or minus one inch[3]. Although age estimates are difficult, it seems likely that he was between thirty and forty-five years of age, based on his hair and beard development and his general physical appearance[4].

The reason why the Shroud has both back and front images of the man is that it would seem that he was placed on his back on one end of the material, with his head towards the centre of the cloth, and the other end was then pulled over his head and drawn up over his front. The front and back images are different in length, with the back image being the longer. Recent research has suggested that this is due to the double curve of the body. His legs were bent and his head appears to have been bent forward[5].

The man has more than 100 dumbell-shaped welts on him, consistent with having been caused by a weighted whip. There also appear to be bloodstains from wounds in his hands, feet, chest and head. Particularly noticeable is a reverse "figure-3" pattern in the centre of the forehead, which would appear to be a significant flow of blood. Although the figure is faint, there is certainly enough detail for it to have been recognized over the centuries as being similar in all respects to descriptions of the crucified Christ.

It was only in 1898, when the first photograph of the Shroud was taken by Secondo Pia, that the full detail of the man on the Shroud was seen in the negative produced by Pia. Relief and depth could be seen and the full horror of the injuries to the man on the Shroud could be realized. Quite naturally, these injuries attracted the attention of medical researchers and experts. The first medical study of Pia's photographs was carried out by a team at the Sorbonne in Paris in 1900, led by a biologist, Paul Vignon. Their findings were presented by one of the team members, Prof Yves Delage, in a lecture to the Paris Academy of Sciences. He explained that, from a medical point of view, the wounds and other markings on the Shroud were so anatomically flawless that they could not have been the work of an artist. It was his conclusion that the Shroud bore the image of Christ, created by some unknown process as he lay in the tomb. His views were considered so controversial by the Academy that it refused to publish the full text of his lecture[6].

The photographs taken by Giuseppe Enrie in 1931 led to further medical interest. A study of the Shroud and research on the medical aspects of crucifixion were carried out by Dr Pierre Barbet of Paris in the 1930s. Dr Barbet noted the wounds of the scourging, of the crowning with thorns, of ill-usage that had taken place, the carrying of the cross, the crucifixion and the blow of the lance[7]. This was one of the early references to ill-usage of the man on the Shroud. The evidence of assault to the facial region of the man on the Shroud had not been noticeable until the photographic negatives had been studied. Dr Barbet refers in particular to a swelling in the right zygomatic region (this refers to the right side of the forehead) and a fracture of the posterior nasal cartilage[8].

Dr Barbet also discusses the crown of thorns in some detail. He points out that none of the evangelists described the shape of the crown of thorns. Traditional Christian art, dating from the fifteenth century, has shown the crown as a circular band of intertwined thorns. However, Dr Barbet quotes two Christian mystics, St Vincent of Lerins and St Brigit, as describing the crown of thorns as having covered the whole of Jesus's head. The man on the Shroud has wounds covering his entire cranium. There is a large accumulation of blood behind the head, consistent with the crown coming into repeated contact with a solid object behind the head. The flows in front are milder but easier to see. In particular there is one major flow that starts high up on the man's head, by the hairline, and flows down in a crooked manner, broadening as if it had met obstacles[9]. In an Appendix to Dr Barbet's book, *A Doctor at Calvary*, Dr P J Smith uses the term "a cap" rather than a crown of thorns, to describe how the crown covered the whole surface of the man's head[10].

In the 1960s an English doctor, Dr David Willis, started to gather and evaluate all the medical research on the Shroud to date. He listed seven facial injuries – swelling of both eyebrows; torn right eyelid; large swelling below the right eye; swollen nose; triangular-shaped wound on the right cheek; swelling on the left cheek; swelling to the left side of the chin[11]. These injuries are all totally consistent with the Gospel descriptions of how Jesus was struck in the face, both at the house of Caiaphas and later by the Roman soldiers.

Dr Willis also refers to "something like a cap of thorns" being responsible for the numerous puncture wounds on the scalp of the man on the Shroud. He describes the blood flows from these wounds and suggests that the different directions of these flows suggests a tilting of the head at various times during the wearing of the cap[12]. He also notes the major flow on the forehead, which he describes as being in the shape of a reversed three. Dr Willis's opinion is that the obstruction in the downward path of the blood flow was due to a reflex contraction of the muscles of the brow, resulting from the pain of the wounds, causing a furrow on the surface of the skin.

Dr Barbet and Dr Willis also refer to injuries to the man's knees and shoulder. Dr Barbet refers to apparent injuries that seem like grazes to both the left and right knees, quoting studies carried out by Dr Giovanni Judica-Cordiglia of the University of Milan[13]. The damage to the right knee seems worse than that to the left knee, as would be expected from a right-handed man when falling – he would fall first on to his right knee. These grazes are quite consistent with the injuries a man would receive if he was to stumble and fall on uneven or rocky ground. Jesus would have stumbled on numerous occasions – between Gethsemane and the house of Caiaphas, and again on the path to Calvary.

Dr Barbet also refers to a broad area of grazing – he uses the term *excoriation* – on the right shoulder of the man on the Shroud. There is also a smaller graze at the point where the man's shoulder blade would be. Dr Barbet interprets this as having been caused by some heavy weight carried by the man on his shoulder. This weight would have had a rough surface that would have damaged the skin of the shoulder through the man's shirt or tunic. If he stumbled while carrying this weight, it could swing round and injure the man on the left side of his back[14]. Again these injuries are totally consistent with a man carrying a wooden beam, such as a *patibulum*, on his right shoulder and stumbling under its weight. Dr Willis, in his notes, reaches the same conclusions as Dr Barbet[15].

As suggested in St Luke's Gospel, Jesus probably did not carry his *patibulum* far before being assisted by Simon of Cyrene. However, even a short distance would have been sufficient for him to incur injuries to his shoulders and knees. In his weakened condition he is likely to have stumbled on the uneven road surface and fallen under the weight of his burden within minutes of taking it up.

* * * * * * * *

The back, buttocks, legs, chest and abdomen of the man on the Shroud bear a large number of dumbbell-shaped markings, consistent with the man having been scourged with a Roman *flagrum* with weights knotted into the ends of the thongs. There are no scourge marks on the arms, which is consistent with the arms having been elevated above the shoulders at the time of the scourging. Dr Frederick Zugibe, noting that some of the marks are difficult to discern, says that he was able to count between 98 and 105 such marks[16]. As the *flagrum* consisted of at least three thongs, each lash would have caused at three lash marks. Dividing 105 marks by three gives a total number of lashes of 35. Again, this is consistent with the Jewish prohibition of not permitting more than 39 lashes.

Research on the pattern of the scourge-marks has suggested that they were inflicted from two directions. This would suggest two scourgers, on opposite sides of the victim[17]. The centre from which the blows were struck on the right side appears to be a little higher than that on the left, suggesting that the scourger on the right of the victim was taller than the one on the left. The scourger on the right also seems to have had a sadistic tendency to lash more at the victim's legs, judging from the pattern of the scourge marks on the back and legs of the man on the Shroud[18].

There is a very clear wound imprint on the left hand of the man on the Shroud. This is the cause of some controversy. There are two streams of blood from the wound which require an explanation. Until very recently, scientists, doctors and writers studying the Shroud have accepted the theory of Dr Pierre Barbet that crucifixion causes asphyxiation, and that the victim seeks to relieve the effects of this by raising and lowering himself on the cross. Dr Barbet and those who have followed him have therefore explained the dual blood flow by suggesting that the arms of the man on the Shroud were in two different positions while he was on the cross, resulting in two separate directions of blood flow. Ian Wilson has estimated from the directions of the blood flows that the arms of the man on the Shroud must have been raised at positions varying between fifty-five and sixty-five degrees from the vertical – a crucifixion position. At sixty-five degrees the body of the victim would have been fully suspended and at fifty-five degrees he would have raised himself in order to breathe more easily[19].

Recently, however, Dr Frederick Zugibe's crucifixion studies have shown that the position of a *crucarius* on the cross does not cause asphyxiation. He specifically reported that every volunteer he used in his studies "affirmed that he had absolutely no trouble breathing during either inspiration or expiration, in itself disproving the asphyxiation hypothesis."[20] In addition, the volunteers were totally unable to lift or straighten themselves while on the cross.

What, therefore, was the cause of the dual blood flow from the wrist of the man on the Shroud? Dr Zugibe's suggestion is that the blood flow occurred when the nail was removed from the wrist, when the body of the victim was taken down from the cross. Dr Zugibe has noted that, as a medical examiner, he has frequently observed blood flowing or oozing from wounds some time after death. Violent death results in increased blood fluidity[21]. As the nail was removed, blood would have flowed out, separating into two streams around the ulnar styloid protuberance (the boney bump on the back of the wrist on the little finger side)[22].

Dr Zugibe places the wound in the region behind the palm of the left hand of the man on the Shroud as being where the wrist meets the metacarpal bones of the back of the hand. This corresponds with a nail being driven through the threnar fissure of the palm rather than through the Space of Destot in the wrist, again contradicting the theories of Dr Barbet[23].

Another feature of the Shroud that has been remarked on by many writers is the fact that no thumbs are visible on the hands of the man on the Shroud. Dr Barbet suggested that when the nail was driven through the Space of Destot it invariably damaged the median nerve. This would have resulted in the contraction of the threnar muscles, causing the thumb to bend sharply inwards towards the palm[24]. Dr Zugibe disputes this and quotes hand reconstruction surgeons as agreeing that, contrary to Dr Barbet's theory, damage to the median nerve might cause an initial rapid inward flexing, but this would be followed by an extension of the thumb. Nailing in the manner suggested by Dr Barbet would therefore have left the thumbs visible. Dr Zugibe's explanation for the missing thumbs is far simpler – the natural position of the thumb, both in death and in a living person, is in the front of and slightly to the side of the index finger. In the case of the man on the Shroud, his palms and index fingers would have been between his thumbs and the surface of the Shroud, preventing any contact with the Shroud[25].

The man on the Shroud has a major wound in his right chest, corresponding to the spear thrust to the dead Jesus described by St John in his Gospel. The wound is almost oval-shaped and is in the space between the fifth and sixth ribs[26]. The size and shape of the wound match excavated examples of the Roman *lancea*, a short spear[27]. The stain from the wound is comprised of both blood and non-blood components, indicating the presence of fluid in the chest cavity, which was released by the spear. The release of this fluid was followed by blood flowing as a result of the heart being pierced[28]. This is, once more, totally consistent with the descriptions of the crucifixion of Jesus in the Gospels and with Dr Zugibe's medical explanation of the cause of the flow of "blood and water".

* * * * * * * * *

Numerous other medical experts have examined the Shroud of Turin over the last fifty years. One of these was Dr Robert Bucklin, a forensic pathologist in Nevada, USA. Dr Bucklin approached the problem of the man on the Shroud as he would have done a standard autopsy. His conclusions matched those of other all the other medical studies and he concluded that the man on the Shroud was dead, having undergone puncture injuries to his wrists and feet, puncture injuries to his head, multiple traumatic whip-like injuries to his back and post-mortem puncture injury to his chest area, releasing both blood and a watery type of fluid. He went one step further. He stated that "it is not an unreasonable conclusion for the forensic pathologist to determine that only one person historically has undergone this sequence of events. That person is Jesus Christ."[29]

It is quite clear that the characteristics of the image on the Shroud match, without any exception, every aspect of the Passion and death of Jesus Christ as described in the Gospels and by forensic experts. Of course that does not prove in scientific terms that the Shroud is indeed the shroud of Jesus. However, it is part of a body of evidence that points in a single direction.

* * * * * * * * *

Another startling but controversial piece of evidence in favour of a first-century date for the crucifixion of the man in the Shroud has been the alleged identification of coin images over the eyes of the man.

In 1976 two United States Air Force physicists, Dr John Jackson and Dr Eric Jumper, placed a photograph of the Shroud into an instrument that had recently been developed, known as the Interpretation Systems VP-8 Image Analyzer. This instrument had been developed as a result of research by the National Aeronautics and Space Administration (NASA) as part of the American space programme. This instrument has the capability of translating light and shade on a black and white photograph into relief. Drs Jackson and Jumper became the first to discover that the Shroud contained relief (three-dimensional) information of the body it had once contained. One of the results of the investigations that they carried out, together with another researcher, Rev Kenneth Stevenson, was that they found small disc-like objects over the eyes of the man on the Shroud. They theorized that the discs might be coins and contacted Ian Wilson on this discovery. It was Wilson who suggested that if the discs were indeed coins, they were the right size to be *leptons*, the widow's mite referred to in the New Testament. Between 29 and 31 AD Pontius Pilate had minted *leptons* in Judaea that lacked the image of Caesar, a factor that would have recommended their use to orthodox Jews[30].

In 1979 a Jesuit priest and professor of theology, Fr Francis Filas, photographed an enlargement of the face of the man on the Shroud. He also noticed some sort of design over the right eye and showed this to a Greek classical numismatist in Chicago, Michael Marx. Marx identified four curving capital letters on the design: **UCAI**. Further research showed them that the objects on the eyes matched in size and shape a coin of Pontius Pilate. The letters that they read as **UCAI** occurred in the correct position for a *lepton* of Pilate.

On the Pilate *lepton* the words TIBERIOU **KAI**SAROS occur, surrounding an astrologer's staff or *lituus*. Fr Filas claimed to also identify a *lituus* on the Shroud image. He suggested that the letters **UCAI** that had been identified were the four letters in the name of Tiberius, but with the **K** substituted by a **C**. It is interesting to note that the *lituus* was an image used in coins minted by Pontius Pilate between 29 and 32 AD, but never by any other official in Palestine or indeed in any other part of the Roman Empire. This is clearly very date and place specific. At the time, however, there was no evidence that Pilate had ever minted a coin using the letter **C** rather than **K** in the word Caesar. Later that same year, however, Fr Filas came across a Pilate *lepton* with the mis-spelling using the letter **C**, thereby proving that his hypothesis was at least feasible[31].

Fr Filas submitted a Pilate *lepton* and his Shroud image to the Spatial Data Analysis Laboratory at the Virginia State University for comparative analysis. This analsys was carried out by Dr Robert Haralick, who provided cautious support for Fr Filas's hypothesis. In his report Dr Haralick said:

"(The) evidence cannot be said to be conclusive evidence that an image of the Pontius Pilate coin appears in the right eye of the Enrie Shroud image… however, the evidence is definitely supporting evidence because there is some degree of match between what one would expect to find if the

Shroud did indeed contain a faint image of the Pilate coin and what we can in fact observe in the original and in the digitally produced images."[32]

Another researcher, Dr Alan Whanger, in 1981 developed a technique known as polarized image overlay technique, which he used to compare various images. He stated that he had identified the coin over the left eye of the man on the Shroud as a "Julia" *lepton*, minted only in 29 AD by Pontius Pilate (and named after Julia, the mother of the Emperor Tiberius). He claimed to find 73 points of congruence between the image on the Shroud and a Julia *lepton* in an area smaller than a finger print[33].

This identification of a Julia *lepton* was also made by two Italian researchers, Professor Pierluigi Baima-Bollone and Professor Nello Balossino, of the University of Turin[34].

However, these claims have been questioned by other researchers on two grounds. Firstly they dispute that it was a Jewish burial custom to place coins over the eyes of a corpse. Secondly, they dispute the identification itself of the two coins.

St John has told us in his Gospel that Jesus was buried in accordance with Jewish burial customs. In the last 30 years there have been a number of archaeological discoveries connecting coins to Jewish burials of the first and second centuries. The usual Jewish burial practice was to lay out the deceased in a shroud or coffin for about a year, during which time the body would decompose. After this time the bones would be collected and placed in an ossuary – a large, rectangular chest or container frequently made of limestone. Obviously this practice of secondary burial has made it difficult to identify any coins found in ossuaries with a specific practice of placing them over the eyes of the deceased. However in 1970 a buried man was found at a fortress in the Jewish desert with silver coins from about 133 AD placed over both his eye sockets. In 1979 Prof Rachel Hachlili described having found coins in skulls in tombs dating from the first century BC to the first century AD. She stated that "the coins originally must have been placed on the eyes of the deceased" and added that this practice was followed often[35]. However, she subsequently changed her mind to suggest that the Jericho coins had been placed in the mouths of the deceased, rather than on their eyes[36]. She has also been quoted as saying that the tombs in question were in bad condition. The ossuary was full of piled-up bones and it was no longer intact[37].

However, experiments have shown that it is possible for coins to fall into the skulls through the upper eye sockets only and not through the mouth[38].

Another Israeli scholar, Zvi Greenhut of the Israel Antiquities Authority, has stated:

"I believe we must now regard coins discovered in the context of Jewish tombs from the Second Temple Period to be elements connected to the burial ceremony, despite the fact that they have not always been found in direct relation to the skulls or bodies of the deceased."[39]

Dr Antonio Lombatti of Italy has claimed that Tiberius's coins always had the letters **OY** rather than **OU** and that the Greek word for Caesar on these coins always started with **K** rather than **C**. He suggested that the reading of a **C** on some coins was the result of degradation or corrosion of the coin rather than being the original spelling. He went on to quote a specialist in Judaic cemeteries, Prof L Y Rahmani, the Director of the Jerusalem Museums, as rejecting without hesitation the idea of a Judaic custom of putting coins on the eyelids of the dead[40].

A technical photographer who has studied the Shroud in great detail, Barrie Schwortz, concluded after studying high quality negatives of the Shroud taken in 1978:

"My personal opinion, based on my photographic experience and my close examination of the Shroud itself, is that the weave of the cloth is far too coarse to resolve the rather subtle and very tiny inscription on a dime-sized ancient coin…. What (Fr Filas) saw as inscriptions, I saw as random shapes and noises. Such is the subjective nature of image analysis. For these reasons, however, I cannot accept these coin "inscriptions" as viable evidence of a first century Shroud 'date'."[41]

An imaging expert from NASA's Jet Propulsion Laboratory in Pasadena, Don Lynn, who was also a member of the Shroud of Turin Research Project (STURP) team in 1978, also believes that the weave of the Shroud is too coarse to resolve the fine inscriptions on a coin the size of a *lepton*. The way he put it was that the pixel size is too large to resolve such fine detail. He concluded that much of what Fr Filas claimed as inscriptions were anomalous weave artifacts caused by five generations of enlarging, copying and contrast enhancing the image using orthochromatic (high contrast) film. He worked from close-ups of the eyes made on ultra-fine grain, ultra high resolution panchromatic (continuous tone) black and white film which had been shot in 1978[42].

Undisputed evidence of Pontius Pilate coins on the Shroud would be very significant in dating it to first-century Jerusalem. However it can be argued that, all too frequently, people see what they want to see. The best that can be said at present about the theory of the Pontius Pilate *leptons* on the eyes of the man on the Shroud is that it remains "Not Proven".

* * * * * * * * *

Another controversy relating to Jewish burial customs has arisen over the question of whether the man on the Shroud was washed before burial. Ian Wilson has suggested that the body visible on the Shroud was not washed and goes on to suggest that this is consistent with the fact that it was very late in the afternoon, immediately prior to the Sabbath, when Jesus was taken down from the cross and that there would have been no time to wash the body in accordance with Jewish burial practices. Wilson fully accepts that it would have been Jewish custom to have washed the body prior to burial and refers to tradition that says that the body of Jesus was washed, and to scriptural scholars insisting that Jesus was washed[43]. It is merely Wilson's own observation or belief that the man on the Shroud was not washed that leads him into this discussion.

Other writers are more adamant that the body of Jesus could not have been washed. The author Mark Antonacci also refers to the lateness of the hour that Friday afternoon, and the time it would have taken to get Pilate's permission to take the body down from the cross, obtain a shroud and carry out the basic formalities of burial. However, he goes on to quote Jewish scholars as saying that Jewish burial rites would have positively prohibited the washing of a blood-stained body that died under the circumstances in which Jesus died. He quotes various sources as consistently concluding that, "if a man dies a violent death and blood is shed, the blood is not washed from the body."[44]

Another writer, John Iannone, refers to the Code of Jewish Law as having special provision for those who were victims of a bloody and violent death:

"One who fell and died instantly, if his body was bruised and blood flowed from the wound, and there is apprehension that the blood of the soul was absorbed in his clothes, he should not be cleansed, but they should inter him in his garments and boots, but above his garments they should wrap a sheet which is called *sovev*."[45]

It was Dr Frederick Zugibe who made the simple observation that the man of the Shroud would have bled copiously from his wounds and went on to conclude that:

"The body unquestionably would have been literally covered with blood because the heart pumps about 4500 gallons of blood through more than 60,000 miles of large and small blood vessels throughout the whole body each day. Instead of the very exact imprints of the wounds, the Shroud would instead bear large indistinct masses of blood over the entire image including the face, arms, hands, feet and trunk. Every practicing forensic pathologist knows that even tiny wounds may bleed profusely during heart activity and observes the end results of bleeding from wounds of practically every type on a daily basis."[46]

Even Mark Antonacci, who argues strongly that neither the man on the Shroud nor Jesus would have been washed prior to burial, concedes that Jesus would have had blood all over his body, front and back, literally from head to feet[47].

Dr Zugibe also quotes an apocryphal scriptural work, *The Lost Gospel According to Peter*, as specifically stating that Jesus was washed before being placed in his shroud. This work dates to the second century AD and was referred to by early church writers such as Serapion, Origen and Eusebius. Such an early document is most unlikely to make such a clear reference if Jewish custom had indeed forbidden the washing of Jesus's body.

Dr Zugibe has also clarified that it is only blood flowing after death that is regarded as unclean, and only above a specific amount, described in Jewish law as a "quarter-log" (an amount equal to about one and a half eggs[48]). It is Dr Zugibe's opinion that the washing of Jesus's body would not have contravened any Jewish prohibitions on unclean blood and could have been carried out in minutes[49]. Therefore the fact that the man on the Shroud was washed is not inconsistent with the burial procedures carried out for Jesus.

In fact the man on the Shroud could not have been completely washed, as there are clear patches of blood from the wounds in his side, on his hands and feet and on his brow. These are all places where blood would have flowed after death: the wound in the side was inflicted after death; the flows from the hands and feet most likely occurred as the nails were removed when the body was taken down from the cross; and the flows on the head could have occurred as the crown of thorns was removed after death. In other words these specific blood flows would have been considered unclean in Jewish law. Hence those washing the body would have avoided them.

Post-mortem blood must, according to Jewish law, be buried with the deceased. Quite obviously, blood on a shroud will indeed be buried with the deceased. This, however, leads to a further interesting point of speculation. If it is assumed that the Shroud is indeed that of Jesus, who, according to the Gospels, rose from the dead, then what would be the position of Jewish law regarding the post-mortem blood flows of a man who had died and then returned to life? Clearly the apostles would have had a somewhat ambiguous view of whether the shroud of Jesus should be considered clean or unclean. It seems likely that, after the discovery of the empty tomb, the shroud remaining there would have been taken away and kept in a safe place, but probably not examined very closely. If there was some form of image on it, this may well not have been noticed at the time, in view of the undoubted turmoil in the minds of the apostles. The shroud is also most unlikely to have been examined in the immediate aftermath of the Resurrection. It would have been stored away, so any image could have been overlooked for possibly centuries. This could well be the answer to the argument frequently put forward by Shroud sceptics that the Shroud of Turin cannot be the shroud of Jesus as there is no mention in the Gospels of an image on the shroud in the empty tomb.

Chapter 13 – The Sudarium of Oviedo

St John's Gospel describes how, when Peter entered the tomb on the morning of Jesus's resurrection, he saw:

"... the linen cloths on the ground, and also the cloth that had been over his head; this was not with the linen cloths but rolled up in a place by itself."[1]

In the Cathedral of Oviedo, a town in northern Spain, there is a piece of cloth that measures approximately 84 x 53 cm. This cloth has visible stains on it. According to tradition this cloth is the "cloth that had been over his head" that is referred to by St John. As it is therefore a purported relic from the burial of Jesus Christ, the question of whether it has any relationship to the Shroud merits investigation.

The primary source for the history of the Sudarium is a twelfth-century bishop of Oviedo, Pelagius. Pelagius recounts how the Persians under their king, Chosroes II, started a war against the Byzantine Emperor Phocas in the early part of the seventh century The Persians were successful in their campaign and their victories included the capture of Jerusalem in 614. At the time there was in Jerusalem an ark, or chest, containing a large number of so-called relics, which were listed by Pelagius. They included such fanciful items as the Lord's mother's milk, hair of the holy innocents and their knuckles, and the hair with which Martha and Mary dried the Lord's feet. Towards the end of the list mention is made of "the Lord's Sudarium"[2].

According to Pelagius, whose account is contained in a collection of documents known as the *Book of Testaments*, this ark was first moved to Africa by Philip, presbyter of Jerusalem and a companion of the presbyter Jerome. It was subsequently taken to Toledo in Spain by Fulgentius, the bishop of the African church of Rusp. It seemingly arrived in Spain during the rule there of the Gothic king Sisebutus, who was a contemporary of the successor of Phocas, the Emperor Heraclius. Pelagius's account is consistent with known history, except in his reference to the presbyters Philip and Jerome, who in fact lived nearly two centuries before the events he described. The reference to Africa almost certainly refers to Alexandria in Egypt. Alexandria was conquered by Chosroes in 616 and it is therefore likely that any relics taken there for safe keeping would have been moved again[3].

Pelagius goes on to describe how a Moslem army invaded Spain and defeated the last Gothic king Rodrigo – this invasion took place in 711. The Spanish Goths retreated to Asturias and took the ark with them. Some years later another Spanish king, Alfonso II, known as the Chaste or the Younger, won a victory over the Moslem invaders which allowed him to establish his capital at Oviedo. It was at this time that the ark was brought there.

Pelagius has been criticized as a historical source and it is of course easy for the sceptic to reject the whole account just from reading the supposed list of relics contained in the ark. However it is necessary to appraise his work with an open mind. The fanciful claims regarding some of the so-called relics does not automatically discredit the story of the Sudarium.

Pelagius describes the ark as having been made by the "disciples of the apostles". This reference clearly links the ark to the apostles and their followers – possibly the second generation of believers in Jesus. It would seem logical that it was the apostles who removed Jesus's burial cloths from the empty tomb. From the description in St John's Gospel, these included the cloth that had been over his head, referred to as a *soudarion*. There are early traditions associating St Peter with the burial cloths and the Sudarium in particular. The apocryphal Gospel of the Hebrews, which dates to no later than the middle of the second century, makes a curious reference:

"But the Lord, after giving the burial shroud to the priest's servant, went to James and appeared to him."

Some writers have suggested that the term "priest's servant" comes from a corruption of the original text, which supposedly referred specifically to Peter. However, there is little evidence for this[4].

A text that does relate the Sudarium with Peter is the Life of Saint Nino of Georgia, who died in 338. This text probably dates from no later than the fifth century. In it is related the legend that the wife of Pontius Pilate became a Christian and that she found the Shroud and gave it to St Luke, none of which has any basis in fact whatsoever. However, it also relates that the Sudarium was found and taken by St Peter[5]. One point of interest here is the suggestion that the Shroud and the Sudarium did pass into different hands at a very early stage.

Another writer who referred in his writings to the Sudarium was Ishodad of Merv, a ninth century bishop in what is now Turkmenistan in Central Asia. Ishodad wrote *Commentaries on the Gospels*. One of his commentaries was on Chapter 24 of Luke's Gospel and Ishodad makes the comment that "and those (the clothes) that were in his grave were taken away by Simeon and by John".[6] The term "clothes" here is seen as a reference to the burial wrappings used for Jesus, as his original clothes had been taken by the Roman soldiers who crucified him.

Ishodad is more specific in his commentary on St John's Gospel, where he wrote:

"… but they gave the garments and linen clothes to Joseph the Senator; for it was right that they should be returned to him, and be kept for him as the lord of the grave…. But the sudarium Simeon took, and it remained with him, that it might be a crown upon his head. And whenever he made an ordination, he arranged it on his head; and many and frequent helps flowed from it; just as even now leaders and bishops arrange the turbans that are on their heads and about their necks in place of that sudarium."[7]

Simeon, in Ishodad's writings, is usually identified with Simon Peter, but there is another possibility. There was a Symeon who was very prominent in the early church in Jerusalem. He was the son of Clopas, who was in his turn the brother of Joseph. Symeon would therefore have been the cousin of Jesus. He became the leader of the Jerusalem church after the martyrdom of James the brother of the Lord in 62 AD. In Chapter 1 I have suggested that the Shroud possibly remained for a number of years in the possession of Jesus's immediate family. Here is a possible reference to a close relative of Jesus in the context of the Sudarium.

Peter left Jerusalem at a relatively early date, probably in the early 40s AD. It seems clear that he did not take the Sudarium with him. Ishodad's reference seems to indicate that the Simeon who took the Sudarium kept it and used it for ritual purposes for a considerable period of time. This is also inconsistent with the story of Peter having had it.

Ishodad wrote in the ninth century and one may reasonably ask whether his commentaries on the Gospels have any historical reliability or basis at all. Possibly not, but his reference to Simeon remains interesting.

Whatever the truth about its early ownership, the Sudarium seems to have remained in Jerusalem. An anonymous text of about 570 describes a pilgrimage to the Holy Land during which a group of pilgrims visits a cave on the bank of the Jordan which is said to contain "the Sudarium that covered Jesus' head"[8]. There is no reason why the Sudarium could not have survived in Jerusalem for six centuries. The church in Jerusalem survived the destruction of the city by the Roman army in 135. As the church historian Eusebius described, the bishops of Jerusalem after the destruction were Gentile rather than Jewish, and the church would have become Gentile rather than Jewish in nature, but it continued to exist. It was normal practice for relics of the early days of the church to be kept and venerated. Jerusalem remained under Roman and then Byzantine rule until the early seventh century and was the destination for many pilgrims. Relics of Jesus' life, passion and death would have been, and indeed were, the subjects of great devotion and veneration.

* * * * * * * * *

The city of Oviedo was founded in 761. It became the capital of King Alfonso II but was sacked by the Moorish invaders of Spain in 794. It was re-established as Alfonso's capital in 795. Pelagius seems to suggest that the ark containing the Sudarium was moved to Oviedo before 761, but he admits that he used the name "Oviedo" for the region of Asturias in general in describing how the ark stayed for a long time in tents.

According to Pelagius Alfonso considered that he had a divine mission to build a resting place for the relics. Accordingly the Church of San Salvador in Oviedo was built during his reign and the Camara Santa, where the Sudarium is kept today, was built within the grounds of the cathedral[9].

There are a number of references to the relics in Oviedo in tenth and eleventh century Spanish texts although there is no specific mention of the Sudarium. During this time the ark was in fact sealed and there could not have been any first-hand evidence of what it contained. Pelagius himself gathered other chronicles into a work generally referred to as the *corpus pelagium.*

Another early mention of the Sudarium in Oviedo is contained in a document known as Codex 99 in the Municipal Library of Valenciennes in France. This describes the opening of the ark (which took place in 1075) and lists the relics found in it. Specific reference is made to "Part of the Lord's Sudarium"[10]. The list again contains many fanciful items which were clearly seen as miraculous and wonderful in the extreme at the time. The Sudarium was one relatively unimportant item among many amazing ones. It can therefore be understood why a number of lists of the time of the contents of the ark, including lists in the *corpus pelagium,* make no mention of it. While there are differences between the narrative of the ark's history in Valenciennes 99 and Pelagius, the basic history and description of the ark and its contents are the same.

The ark was opened in Oviedo on 14 March 1075, in the presence of King Alfonso VI, his sister and the knight Rodrigo Diaz de Vivar, known as El Cid. A list was made of the relics inside it, which included the Sudarium[11]. The Sudarium has been kept in the cathedral in Oviedo since that time, so its history since then is known without doubt.

* * * * * * * * *

It was first suggested in 1985 by Monsignor Giulio Ricci that the Sudarium of Oviedo and the Shroud of Turin had really been used on the same corpse and that this merited investigation. Studies were subsequently carried out on the Sudarium by the Investigation Team of the Spanish Centre for Sindonology (EDICES). This team was led by an engineer,

The Sudarium of Oviedo

Guillermo Heras Moreno, and included, among others, Professor Jose-Delfin Villalain Blanco, Professor of Forensic Medicine at the University of Valencia, and Professor Jorge-Manuel Rodriguez Almenar, also from the University of Valencia. The American physicist Dr John Jackson also worked with the Investigation Team.

The team carried out forensic, geometrical and mathematical studies, and published the following conclusions at the First International Congress on the Sudarium of Oviedo, held in Valencia in 1996.

1. The Sudarium of Oviedo is a relic, which has been venerated in the cathedral of Oviedo for a very long time. It contains stains formed by human blood of the group AB.
2. The cloth is dirty, creased, torn and burnt in parts, stained and highly contaminated. It does not, however, show signs of fraudulent manipulation.
3. It seems to be a funeral cloth that was probably placed over the head of the corpse of an adult male of normal constitution.
4. The man whose face the Sudarium covered had a beard, moustache and long hair, tied up at the nape of his neck into a ponytail.
5. The man's mouth was closed, his nose was squashed and forced to the right by the pressue of holding the cloth to his face. Both these anatomical elements have been clearly identified on the Sudarium of Oviedo.
6. The man was dead. The mechanism that formed the stains is incompatible with any kind of breathing movement.
7. At the bottom of the back of his head, there is a series of wounds produced in life by some sharp objects. These wounds had bled about an hour before the cloth was placed on top of them.
8. Just about the entire head, shoulders and at least part of the back of the man were covered in blood before being covered by this cloth. This is known because it is impossible to reproduce the stains in the hair, on the forehead and on top of the head with blood from a corpse. It can therefore be stated that the man was wounded before death with something that made his

scalp bleed and produced wounds on his neck, shoulders and upper part of the back.

9. The man suffered a pulmonary oedema as a consequence of the terminal process.

10. The cloth was placed over the head starting from the back, held to the hair by sharp objects. From there it went round the left side of the head to the right cheek, where, for apparently unknown reasons it was folded over on itself, ending up folded like an accordion at the left cheek. It is possible that the cloth was placed like this because the head formed an obstacle and so it was folded over on itself. On placing the cloth in this position, two stained areas can be anatomically observed – one over the "ponytail" and the other over the top of the back. Once the man had died, the corpse stayed in a vertical position for around one hour, and the right arm was raised with the head bent 70 degrees forward and 20 degrees to the right. This is a position compatible with death by crucifixion.

11. The body was then placed on the ground on its right side, with the arms in the same position, and the head still bent to the right, and at 115 degrees from the vertical position. The forehead was placed on a hard surface, and the body was left in this position for approximately one hour.

12. The body was then moved, while somebody's left hand in various positions tried to stem the flow of liquid from the nose and mouth, pressing strongly against them. This movement could have taken about 5 minutes. The cloth was folded over itself all this time. The cloth was then straightened out and wrapped all around the head, like a hood, held on again by sharp objects. This allowed part of the cloth, folded like a cone, to fall over the back. With the head thus covered, the corpse was held up (partly) by a left fist. The cloth was then moved sideways over the face in this position. Thus, once the obstacle (which could have been the hair matted with blood or the head bent to the right) had been removed, the cloth covered the entire head and the corpse was moved for the last time, face down on a closed left fist. This movement produced the large triangular stain, on whose surface the finger shaped stains can be seen, and on the reverse side of the cloth, the curve inscribed on the cheek.

Like the previous movement, this one could have taken five minutes at most.

13. Finally, on reaching the destination, the body was placed face up and, for unknown reasons, the cloth was taken off the head.
14. Possibly myrrh and aloes were then sprinkled over the cloth[12].

It does need to be stressed that the Sudarium contains numerous stains but does not contain the specific image of a human being. The stains on the cloth are perfectly natural and there has never been any suggestion that they might have a supernatural or unexplained origin. It is these stains that were studied and measured in detail by the Investigation Team and which have given rise to the conclusions reached by the team.

From a textile point of view, the Sudarium of Oviedo is linen, perpendicular in warp and weft. This is a pattern called taffeta. The linen is coarser than that of the Shroud and may well have been meant for domestic use.

While it was not Roman practice to cover the faces of crucifixion victims, in Jewish crucifixions it was customary to cover the face of the corpse when it became disfigured. The Sudarium of Oviedo shows that all the area of the head it touched was completely covered in blood. A bleeding corpse, with a disfigured, bloody face and pulmonary oedema liquid coming out of the nose and the mouth would have been a sufficiently horrible sight to warrant the use of a cloth to cover the head, in accordance with the general instructions about blood in the Old Testament and the specific instructions of the Sanhedrin at the time. It would have been required for such a blood-stained cloth to be buried with the corpse[13].

There are clear similarities between the evidence of the Sudarium and the evidence of the Shroud. Both the man in the Shroud and the man in the Sudarium had a beard, moustache and long hair tied into a ponytail. Both were of blood group AB. As has been mentioned in Chapter 1, this is a relatively uncommon blood group globally, but is found more commonly among the "Babylonian" Jews and the Jews of northern Palestine.

Both the man in the Shroud and the man in the Sudarium had been tortured before death and both suffered wounds from sharp objects on the nape of the neck. In both cases the manner of death appears to be consistent with crucifixion, as both died in an upright position with their arms outstretched. Both suffered a pulmonary oedema[14].

The similarities between the two cloths go further. The Investigation Team carried out geometrical studies comparing the bloodstains from the head of the man in the Shroud with those from the head of the man in the Sudarium. They found that the size of the stains is geometrically compatible, as is their relative position on each cloth. Similarly bloodstains on the back of the man also correspond on the two cloths – on the Sudarium they can be found in the lower left and right corners.

They also found a number of similarities between anatomical features of the faces of the man in the Shroud and the man in the Sudarium. These include the noses, which are of similar sizes; the superciliary ridges; swellings on the right sides of the noses; tips of the noses, nostrils and nose-wings; the positions and sizes of the mouths; the chins and the shapes of the beards[15].

All in all, there would seem to be a reasonable probability that the two cloths covered the same body. The definitive proof would of course be DNA tests on the bloodstains on the two cloths. These, however, would not be easy in view of problems arising from the age of the blood samples and the relative lack of cells. Was this body that of Jesus Christ?

There is one disconsonant note in the investigations into the Sudarium. A carbon-14 dating of the Sudarium was carried out in the early 1990s and reported on at the First International Congress on the Sudarium of Oviedo held in Oviedo in 1994. According to this dating the cloth dates from the 7th century. However neither the Investigation team nor the Cathedral of Oviedo has any knowledge of the methods used for this dating nor have they seen any definite laboratory report. The scientist who commissioned the test, Prof Pierluigi Baima-Bollone, an Italian specialist in forensic medicine, himself admitted in his report to the Sudarium Congress that "The result is not easy to interpret due to the well-known difficulties of dating textiles and to the conditions under which the sample was kept from when it was taken (1979) until it came to us some years after (Dr) Frei's death in 1983.... The carbon dating we ordered should be nothing more than a stimulus to more precise investigations under better conditions."[16]

A further carbon-dating of a sample from the Sudarium was carried out in 2007, shortly before the Second International Congress on the Sudarium of Oviedo held in Oviedo in April of that year. Again the sample produced a date of around 700 AD[17]. The laboratory that was used (via the National Museum of Madrid) expressed surprise at this and enquired if the cloth was perhaps contaminated with any oil-based product, as oil is not cleaned by the laboratory processes used before carbon dating. If oil is present on the sample the date produced by carbon dating is likely to reflect the date of contamination. At present Sudarium researchers are unable to explain this dating in the context of the history of the Sudarium being well established, with definite references to its presence in Jerusalem in the fifth and sixth centuries.

From an archaeological point of view, the two cloths of Oviedo and Turin have a number of points in common. From the point of view of textile study there is nothing to eliminate the possibility that they existed together – apart from the disputed or uncertain radiocarbon dating results on both cloths. From the forensic point of view the deaths recorded in each cloth are very similar and coincide with the Gospel descriptions of the crucifixion of Jesus Christ.

It is also worth noting that the Sudarium of Oviedo was taken off the body whose head it had wrapped. It was subsequently kept and venerated. From the account in St John's Gospel we know that Jesus was buried in a shroud and the position where the *soudarion* was found indicates that it had been removed before the Shroud was used[18].

Both the Shroud of Turin nor the Sudarium of Oviedo are connected in their respective traditions with the death and burial of Jesus Christ. They both appear to be connected with the same body. It has also been suggested that the ponytail seen on the man on the Shroud was not a specific hairstyle but rather the result of the hair of the victim being pulled back and sewn to the Sudarium[19]. At the end of the day, however, it comes down to a matter of individual conviction and belief as to whether that body was the body of Jesus Christ. As another author has put it:

"...while it is certainly true that there is no scientific test for Christness, it cannot be scientifically stated that the Sudarium is not the one used on Jesus of Nazareth."[20]

Further scientific studies are being carried out on the Sudarium, including DNA studies on the linen itself, directed at establishing the geographical area from where the linen originated[21].

PART 4 –
THE SHROUD AND SCIENCE

Chapter 14 – The First Investigations

It was not until the advent of photography that scientific examination of the Shroud could begin. Before Secondo Pia took his ground-breaking photographs in 1898, the Shroud was seen purely as a religious artifact. One could either believe or not believe that it was the genuine shroud of Jesus Christ; it was all a matter of faith. The image on it was clearly of miraculous origin. How the miracle had occurred was not a question that any of the faithful asked.

Pia's photographs showed for the first time that science had something to say about the Shroud. By revealing the negativity of the image they clearly demonstrated that scientific procedures – in this case photography – could discover more information about the Shroud. Once this realization had been made the door was open to scientific study and debate.

The first scientists to study the Shroud were medical doctors. The nature of the image and the injuries to the man on the Shroud excited the interest, among others, of a French professor of anatomy, Yves Delage. Delage was an agnostic with a wide range of scientific and other interests who clearly felt that the positive image revealed in the photographs deserved more detailed examination. In 1900 he discussed the photographs with another scientist, a young man called Paul Vignon. Vignon had a knowledge of both biology and art, which enabled him to eventually reach, among other findings, the conclusion that the image could not have been formed by painting.

Vignon and Delage, assisted by three other scientists, carried out their investigation from 1900 until 1902. Their major disadvantage was that they did not have access to the Shroud itself and had to work from Pia's photographs. Despite this, their work was a credit to the limited resources and limited technology of the time. On 21 April 1902 Professor Delage gave a detailed report on the investigation to the French Academy of Science in Paris, one of the world's leading scientific bodies. This was the same venue where Louis Pasteur had announced his discovery of a vaccine for rabies.

Delage announced the findings of his and Vignon's research to the Academy. It was their conclusion that the image on the Shroud was that of a dead human male. They found that the image could not have been painted but was the result of direct contact between the cloth and the body it had enfolded. However, they went further. They identified the body as being that of Jesus Christ and declared that the Shroud was indeed that used in his burial. Delage's final recommendation was that the Italian authorities should be approached to allow more detailed examination of the Shroud.

Delage's conclusions were highly controversial in the prevailing intellectual atmosphere of free thought and rationalism in France at that time. The majority of members of the Academy were religious sceptics to whom the suggestion that the Shroud could be that of Christ was totally unacceptable. The permanent secretary of the Academy, Pierre Berthelot, was a leading sceptic who attempted to prevent Delage from presenting his report to the Academy in the first place. Although over-ruled, he ensured that much of Delage's report was excluded from the journal of the Academy. He was in the forefront of criticism of Delage, who was accused of being a traitor to science and to his own agnosticism. Delage's response to his critics remains an outstanding defence of scientific integrity:

"I recognize Christ as a historical personage, and I see no reason why anyone should be scandalized by the fact that there still exists material traces of his earthly life.... If our proofs have not been received by certain persons as they deserve to be, it is only because a religious question has been needlessly injected into a problem which in itself is purely scientific, with the result that feelings run high and reason has been led astray. If, instead of Christ, there were questions of some person like a Sargon or Achilles, or one of the Pharaohs, no one would have thought of making any objection.... I have been faithful to the true spirit of Science in treating this question, intent only on the truth, not concerned in the least whether the truth would affect the interests of any religious party. There are those, however, who have let themselves be swayed by this consideration and have betrayed the scientific method."[1]

Paul Vignon was the first to attempt to explain scientifically how the image had been formed on the Shroud. His hypothesis of image formation was included in Professor Delage's report to the French Academy of Sciences. Vignon studied the effects of paints and dyes on linen. He worked on the assumption, taken from the Gospel of St John, that there would have been aloes used in the burial process of the man on the Shroud. He postulated that urea present in the sweat and blood degrades after death, giving off ammonia gas that rises from the body and reacts chemically with the aloes placed on the cloth. This, he suggested, would give rise to a brown colouration. He also made the critical observation that the image intensity on the cloth varies inversely with the distance between the cloth and the body. It was his theory that the further away from the cloth the parts of the body were, the further the ammonia vapours had to rise, resulting in a less intensive image of that body part, and vice versa[2].

This hypothesis became known as the Vapourgraph Hypothesis. Subsequent scientific studies, using more advanced technologies, have shown it to be untenable.

Prior to developing his Vapourgraph Hypothesis, Vignon had tried a contact approach to try to explain the image. He used ground-up red chalk, which he applied to his face after putting on a false moustache. He then had assistants gently press a linen cloth containing albumin on his face. The resulting image was severely distorted and Vignon reached the conclusion that such an approach could not explain the image[3].

Vignon's pioneering work established firmly that the single most important scientific question about the Shroud is this – how was the image formed? In more recent years a second question has been asked – how old is the Shroud – but study, and controversy, still continues today about the whole question of image formation.

* * * * * * * * *

Little more scientific study of the Shroud seems to have been carried out until the second photographs of the Shroud were taken by Commander

Giuseppe Enrie at its May 1931 exhibition. These photographs dispelled any lagging doubts there may have been in the scientific and medical communities that Pia's photographs were, in some way, a hoax[4].

Research followed in a number of directions. In 1932 two Italian forensic medicine experts, Dr Romanese and Dr Judica-Cordiglia, together with a chemist, Dr Scotti, took renewed interest in the possibility that the image had been caused by direct contact between the cloth and the body – the Contact Hypothesis. Dr Romanese soaked a corpse with a salt solution resembling sweat and then placed a linen cloth containing aloes over it, obtaining negative images of the body. Dr Judica-Cordiglia placed blood on the corpse, covered it with a linen cloth that had been soaked in a mixture of aloes, turpentine and oil, and then exposed the cloth to hot steam. He succeeded in producing better images, but nothing like the quality of the image on the Shroud[5].

In Paris Dr Pierre Barbet began his experiments to show that the wounds visible on the man in the Shroud are indeed those of a victim of crucifixion. He carried out his anatomical experiments in 1932 and 1935[6].

Paul Vignon continued his own studies of the Shroud into the 1930s. He carried out a study of early portraits of Christ and noted a number of points of similarity between these and the facial image on the Shroud. He described 15 of these, which today are referred to as Vignon markings[7].

St Joseph's Hospital in Paris seems to have been the source of varied research into the Shroud. Not only was Dr Barbet from St Joseph's Hospital, but in 1942 a pharmacist there, Dr Jean Volkringer, observed that when certain plant herbs are pressed in a book for a period of years, they frequently produce finely detailed images of the plants on the pages of the book. These images were negatives of the plants, in the same way as the image on the Shroud appears to be a negative of the man. He suggested that a similar process could have been responsible for the image on the Shroud. There is one major difference, however. The image on the Shroud is only present on the surface tips of the Shroud material. In the Volkringer patterns the image penetrates right into the paper. It also seems that the cellulose in the book paper plays a significant role in the production of the images[8].

Seminars and conferences on the Shroud also began to be held. A Congress on Shroud Studies was held in Turin in 1939, at which some twenty papers were presented. The term *sindonology* for studies of the Shroud also seems to have been introduced at about this time. In 1950 an International Sindonological Congress was held, as part of Holy Year celebrations, in Rome[9]. These Congresses were both organized by a forerunner of the Turin International Centre for the Turin Shroud (known in Italian as the Centro Internationale di Sindonologia), which was founded in 1959 to promote scientific studies of the Shroud. The first Director of the Centre was Prof Giovanni Judica-Cordiglia, who had conducted research into the Shroud in the 1930s.

Until the late 1960s, all research on the Shroud had to be conducted from photographs. The custodians of the Shroud in Turin remained adamant in their refusal to allow any scientific access to the Shroud. The possibility of taking physical samples would have been absolute anathema. A breakthrough was achieved in 1969, when a preliminary scientific examination of the Shroud was permitted by the Archbishop of Turin, Cardinal Michele Pellegrino. He permitted the formation of a commission to check the state of preservation of the Shroud and to recommend any suitable scientific tests they felt should be carried out[10]. This is not to say that earlier researchers had never seen the Shroud. In 1933 at the Holy Year exhibition of the Shroud, Pierre Barbet had viewed it from a distance of less than a yard. Paul Vignon also saw it at the same exhibition[11]. However, viewing the Shroud at an exhibition was a far cry from making a scientific examination of it.

The commission set up by Cardinal Pellegrino consisted of ten men and one woman. They were all Italians from the region of Turin and its surrounds. Three of them were priests and five were scientists. Of the remaining two one was a historian and the other was a former director of art galleries. Although from a limited area geographically, they were more than competent intellectually and were selected foremost for their particular academic expertise. Prof Judica-Cordiglia of the Turin International Centre for the Turin Shroud was one of the members of this commission.

The commission examined the Shroud over two days during June 1969, during which time the first colour photographs of the Shroud were taken by Prof Judica-Cordiglia. A report was prepared noting the Shroud's excellent state of preservation as well as recommending a series of tests for a future date. Finally, the commission emphasized that "minimal samples" should be removed from the Shroud[12].

Further studies took place in 1973. Cardinal Pellegrino set up a new commission of experts, commonly referred to as the Turin Scientific Commission. In November that year the Shroud was examined by this commission and, for the first time, samples were taken from the Shroud. Prof Gilbert Raes of the Ghent Institute of Textile Technology in Belgium took a small sample from one end of the Shroud's frontal end, and another small sample from the Shroud's side-strip. He also took one warp thread and one weft thread[13]. Prof Raes's samples were to play a critical role years later in studies on the age of the Shroud.

Prof Raes was able to confirm from his studies that the weave and fabric of the linen of the Shroud are of a type known to have existed in first-century Judaea. He also found traces of cotton among the linen fibres. To be precise, he found that one part of his sample contained cotton, but the part on the other side of a seam did not. He identified that cotton as an ancient Near Eastern variety, *Gossypium herbaceum*[14] and on the basis of this suggested that the Shroud had been woven on a loom that had also been used for cotton. Cotton does not grow in Europe but is abundant in the Near and Middle East[15].

The members of the commission also studied the image itself. They found no pigment or dye in the image areas and their study also showed that the image was superficial – it lay only on the topmost fibres of the threads of the Shroud. They described the image as being composed of "yellow fibrils". The yellow colour did not soak into or penetrate the fibres as a pigment or dye would have done. This observation also seemed to rule out Paul Vignon's Vapourgraph Hypothesis – ammonia-soaked sweat coming out of the body would have penetrated the fibres of the cloth[16].

Dr Max Frei, a Swiss criminologist, took dust samples from between the threads of the Shroud by means of adhesive tape. He took a total of twelve samples in order to try to identify pollen grains that might give some clue to the place of origin of the Shroud. Dr Frei's background as an analytical scientist was impeccable. He had served as president of the United Nations' fact-finding committee to investigate the death of the U.N. Secretary-General, Dag Hammarskjold, in a plane crash in what was then Northern Rhodesia (now Zambia) in the early 1960s. He founded the scientific department of the Zurich Criminal Police and directed it for twenty-five years. He held a doctorate in botany and was an acknowledged expert on Mediterranean flora. Pollen analysis was his special field.

Over a period of nine years he made seven trips to the Middle East to identify the pollens that he had taken in his samples. From the samples he took in 1973 and further samples taken by him in 1978 he identified a total of fifty-eight pollens. Of these, forty-five came from plants found in Jerusalem and its environs. These are generally desert plants, some of which are found frequently in specific areas such as around the Dead Sea, east of Jerusalem, the Jordan valley and Jerusalem itself. Eighteen of the pollens came from the Anatolia region of Turkey, where Edessa was located, and thirteen were also common to Constantinople. Of these thirteen, three are not found in Anatolia or Israel. There were other pollens found, not surprisingly, in France and the Mediterranean area.

Of the non-European species, there is only one place where all of them, with the exception of three, grow in a very small radius – Jerusalem. This appears to be strong evidence that the Shroud originated in the Jerusalem area. Of the three not found in Jerusalem, two were found by Dr Frei in Urfa, Anatolia, the ancient Edessa, and one in Constantinople.

Dr Frei's work and conclusions have, inevitably, been the subject of criticism and scorn. It has, for example, been suggested that the pollens found on the Shroud could have been carried to Europe from the Middle East on the wind. There are several objections to this. Firstly, the Mediterranean wind system is very complicated and a transport of pollens from the Middle East to France or Italy must be considered highly improbable. Secondly, during its history in Europe the Shroud has only been exposed to the open air on limited occasions. It would be amazing if these expositions coincided with precise storms that brought pollen from the Middle East. Finally, the pollens on the Shroud come from plants that bloom in different seasons of the year. Therefore this improbable accident of exposure would have had to have occurred repeatedly[17].

Dr Frei himself has been the subject of personal criticism. Dr John Heller, a member of the Shroud of Turin Research Project (STURP) of 1978, subsequently described Dr Frei as an amateurish criminologist. His background in criminology does, however, belie this accusation. It has also been pointed out that he was one of three experts who authenticated the forged "Hitler Diaries" that were published in 1983, and that this casts doubt on his scientific abilities[18]. It should be noted that in this instance Dr Frei was called as a handwriting expert. His greater expertise lay in pollen analysis and the fact that he made one mistake, however unfortunate, cannot and does not disqualify him from being acknowledged as an expert in his own chosen field.

It has also been noted that the 1978 STURP team claimed to have found little pollen on the samples that they took[19]. However, they used a different method of taking cloth samples with sticky tape. They used a torque applicator, which limited the pressure on the Shroud. Dr Frei did not. As a result the STURP team took samples from the surface of the Shroud only. Dr Frei was able to lift material from between the cloth threads[20].

The pollen evidence by itself does not prove that the Shroud originated in Jerusalem or traveled to Edessa or Constantinople. It certainly does not give a date for the Shroud's manufacture. It is, however, part of a body of evidence that would grow, indicating a Middle Eastern origin for the Shroud. Dr Frei himself was a highly regarded botanical expert who spent fifty years studying the flora of the Mediterranean and the Middle East.

* * * * * * * * *

Until the 1970s, research into the Shroud had been very much the preserve of European scientists. Now American scientists too developed an interest in studying the Shroud. In 1974 two United States Air Force physicists, Dr John Jackson and Dr Eric Jumper, started a spare time research programme on the Shroud. Their initial objective was to investigate the observation of Paul Vignon that the image on the Shroud appeared to be present even where the cloth could not have touched a body. They constructed a full-scale cloth model of the Shroud, based on photographs of the Shroud, and used this to wrap human subjects inside the cloth.

With the subject enshrouded, they were able to measure the distance of various portions of the image on the cloth from the subject beneath. In this way they obtained hard evidence supporting Vignon's observation that the intensity of the image on the Shroud varies inversely with the distance between the cloth and the body[21].

In the course of their research they were introduced by a fellow scientist, Bill Mottern, to the recently-developed Interpretation Systems VP-8 Image Analyser. Using this they were not only able to reinforce their findings regarding the distance relationship, they were also able to generate a three-dimensional image of the man in the Shroud. The three-dimensional image is clear and undistorted, giving for the first time a clear impression of the features of the man in the Shroud. It was from these images that they noted the objects on the eyes of the man that they identified as coins and which were discussed in Chapter 12[22].

Dr Jackson went on to become one of the leading experts on the subject of the physics of the image on the Shroud and its formation, as well as the driving force behind the Shroud of Turin Research Project of 1978.

The observations of the 1973 Commission regarding the nature of the image were analysed by Dr Ray Rogers, a chemist then employed at the Los Alamos National Laboratory in the United States. In his turn Dr Rogers was to become one of the leading experts on the subject of the chemistry of the image on the Shroud and its formation.

Dr Rogers looked at the image in the context of the fire of 1532, which had burned part of the Shroud. He observed that if there had been organic molecules, such as those of pigments or dyes, on the cloth, they would have changed colour or burned at different rates during the fire, depending on the intensity of the heat on different areas of the Shroud. He noted that the parts of the image that were in contact with the burned areas have apparently "*identical* colour tone and density as parts of the image at the maximum distance from a discoloured area". This indicated to him that it was highly unlikely that an organic pigment was responsible for the image.

He also felt that this ruled out the possibility that the image had been formed by direct contact with a body. Such an image would have been formed by organic molecules and hence there would have been variations in the colour of the image due to the 1532 fire.

The fact that there had been no absorption of the image substance into the fibres of the threads at a microscopic level also seemed to rule out the use of any inorganic compunds, such as ink, in the formation of the image[23].

* * * * * * * * *

In March 1977, American scientists held the First US Conference of Research on the Shroud in Albuquerque, New Mexico. A number of papers on the Shroud were presented at this Conference, including papers by Dr Jackson, Dr Jumper and Dr Rogers.

One theory that was discussed in some detail at this Conference was that the image on the Shroud had somehow been the result of some kind of scorch. This hypothesis had first been put forward by a British author, Geoffrey Ashe, in 1966. To the naked eye, the experimental scorch of a piece of linen resembles the colour of the image on the Shroud.

Dr Jackson presented evidence that there was no difference between the colour of the image and the colour of areas of the Shroud that had been burned and scorched in the fire of 1532. To ascertain this he had used a microdensitometer, an instrument that measures the densities of images on photographic film or plate.

Dr Rogers referred to the relationship between the image density and the distance between the body and the cloth, which had been confirmed by Dr Jackson and Dr Jumper the previous year. He said that this relationship could suggest "rapid heating as the cause of the image". He felt at that time that if future testing did not identify any pigment on the cloth, and if no one found an organic stain that could have stained the cloth naturally, then the scorch theory was the only hypothesis left.

Dr Jumper suggested that if the Shroud had been scorched, it would have had to have been the result of a very short burst of high energy radiation. He detailed experiments he had conducted with Dr Jackson showing

The First Investigations

that the radiation process would have had to be very quick and very intense to scorch only the topmost layer of the linen fibres.

Dr Jackson also went a stage further. Through mathematical analysis he showed that no reasonable physical mechanism could produce an image that was both three-dimensional and highly detailed. This seemed to rule out any natural process involving the diffusion of chemicals, but it also seemed to rule out the production of a simple scorch from heat. However, he felt that a scorch could still be the result of some form of radiation other than heat[24].

Another paper was presented by two scientists at the Jet Propulsion Laboratory in Pasadena, Jean Lorre and Donald Lynn. They had analysed the Shroud image with some of the computer-assisted techniques that they had also used to study images transmitted from the surface of Mars by the Viking spacecraft in 1976. They found no "directionality" to the image – in other words it had been applied to the surface of the cloth in a random and directionless fashion. This was another piece of evidence against the image on the Shroud having been applied by a painter. Any application by human hands would show a characteristic pattern, no matter how carefully the artist worked. No such pattern exists on the Shroud.

The implications to the scientists began to look more than substantial. It had been established that the Shroud image was a negative, it was superficial and it contained three-dimensional information. There seemed to be no evidence of any organic substance in the image. They could not imagine how such an image could be produced in the twentieth century, let alone the fourteenth[25].

At this point it seemed clear that further, intensive research on the Shroud was needed. Research on the Shroud had moved to a new level. The latest technologies were being used in the most advanced laboratories in the world to study the Shroud, and the more studies that were conducted, the more questions were raised and the more enigmatic the image seemed to be. The Shroud was clearly more complex than a simple mediaeval forgery. It had now attracted the professional interest of a group of some forty respected scientists from a number of different institutions. They had begun their studies of the Shroud in their own time. Now they would plan

Mark Oxley

to carry out a series of tests on the Shroud itself in Turin to try to answer the questions that they themselves had raised.

Chapter 15 –
The Shroud of Turin Research Project (STURP)

The organization which opened the way for a further scientific examination of the Shroud in Turin was the Holy Shroud Guild, a New York-based organization led at the time by two Catholic priests, Fr Adam Otterbein and Fr Peter Rinaldi. The Guild was formed in 1951 with the objectives of propagating knowledge about the Shroud and to support learned investigation into the Shroud. Fr Otterbein had assisted Dr John Jackson by providing him with clear photographs of the Shroud that had then been used by Dr Jackson in his VP-8 image analyzer studies.

Fr Otterbein and Fr Rinaldi harboured a dream that a scientific study of the Shroud would be permitted at the time of its planned exhibition in 1978, which would mark the 400th anniversary of its arrival in Turin. They had nothing on which to base this dream, except hope. Then, through his relationship with Dr Jackson, Fr Otterbein came to realize that he had found a group of scientists who were seriously interested in a scholarly investigation of the Shroud.

Fr Rinaldi attended the First US Conference of Research on the Shroud in Albuquerque in March 1977 and put the idea there to Dr Jackson of the possibility of a "hands-on" examination of the Shroud the following year. Towards this end the papers presented at the conference were printed and bound in a 243-page book entitled *Proceedings of the US Conference on the Shroud of Turin*, which was then taken to Turin by Dr Jackson and a group of his colleagues for presentation to the Church authorities there. In early 1978 Fr Rinaldi was able to inform Dr Jackson that qualified approval had been given for a scientific examination of the Shroud[1].

One key figure in authorizing the examination was ex-King Umberto II of Italy, then the legal owner of the Shroud. He had favoured an examination of the Shroud for some years and his approval, from his home in exile in Portugal, was forthcoming. But even his approval would have meant little without the support of the Archbishop of Turin. The Archbishop at the time, Cardinal Anastasio Ballestrero, also gave his approval for non-destructive tests to be carried out on the Shroud[2].

The next stage was to obtain money and equipment for the investigation. The chief fund-raiser was Thomas D'Muhlala, the president of Nuclear Technology Corporation in Connecticut. He had contacted Dr Jackson, curious about the proposed examination, and had promptly been invited to attend the next team meeting. He then succeeded in procuring, through gifts and loans, nearly US$ 2,5 million worth of scientific equipment together with cash for travel and other expenses for the team. Members of the team also raised money through selling assets and taking out personal loans[3]. It was a team effort based on mutual commitment.

The participants in the Albuquerque conference comprised the core of the eventual Shroud of Turin Research Project (STURP), as it came to be known in 1978. Other participants, such as Thomas D'Muhlala, joined the team in the intervening period until eventually a group of twenty-seven scientists[4], together with technicians and other support people, was ready to travel to Turin in October 1978.

The STURP team learned at first hand that, however effective one's planning might be, bureaucracy can always create unforeseen difficulties. The crates containing all the scientific test equipment were consigned to the care of Fr Rinaldi in Turin. He had been born in Turin and it was felt that this would be a convenient way of handling the consignment. Unfortunately Fr Rinaldi had not been back in Turin long enough at the time to qualify as being Turinese. As a result he was informed that the equipment would have to be impounded in Milan for 90 days before it could be released. After much argument Fr Rinaldi finally made a personal call to the Minister of Commerce in Rome and secured the release of the equipment from Milan. It was then sent to the Customs Department in Turin, who demanded a substantial bond to be posted before they would release the equipment. Cardinal Ballestrero came to the rescue this time and guaranteed the bond from Turin Cathedral funds. At last the equipment was released[5].

The team had originally been told that they would have only 24 hours on 9 October 1978 to carry out their examination of the Shroud[6]. This was extended to 120 hours, after they had arrived in Turin and while they were waiting for the equipment to be released. On the appointed day, eight tonnes of equipment was delivered to the royal palace of the House of Savoy, next door to the Cathedral, to be met with the news that there was no forklift truck available to offload it. The team carried the equipment into the palace and up two flights of stairs to the work area by hand. The equipment was then set up in 48 hours, time having to be made up because of the delays in having the equipment delivered[7]. Finally the examination was ready to be started on the evening of Sunday 8 October.

The STURP team was joined in their investigation by Dr Max Frei and Italian scientists including Prof Giovanni Riggi and Prof Pierluigi Baima-Bollone who carried out their own independent research programme[8].

* * * * * * * * *

The primary goal of STURP was to examine the physical and chemical characteristics of the image with a view to determining these more exactly[9]. To be more specific, they had three objectives: what is the image composed of; what was the process that formed it; and what is the composition of the bloodstains[10]?

Many of the scientists involved in the examination were initially ready to find that, after all, the Shroud was no more than a very well-crafted painting, in line with the allegations made by Bishop d'Arcis in 1389. As a result many of the tests were designed to address the hypothesis that the image is an artifact and more specifically that it is an applied pigment. Numerous tests were carried out to examine this hypothesis, including direct microscopic examination, different forms of spectrometry, fluorescence studies, photographic imaging, electron microscopy and, finally, samples were taken from the surface of the Shroud for later microscopic and chemical examination. The findings from direct microscopic examination were that the colour of the image does not penetrate the cloth in any image area – it is totally superficial – and there is no evidence for cementation between the fibres or for capillary flow of liquids. No pigment particles could be resolved at 50x magnification. In other words, evidence that would be expected from a painted cloth just did not exist. The team also noted that the image does not look like a painting when studied by direct microscopic examination.

Similarly, other tests revealed not a trace of the dyes, stains, pigments or painting media that would be expected on a painted cloth. In fact, the microscopic and chemical data suggested the image to be the result of some form of cellulose degradation effect rather than an applied pigment[11].

One interesting observation was made during the reflectance spectroscopy examination. The readings were similar over the entire image, as would be expected if the image was made up of a single substance, except at the heel. Microscopic examination of the heel revealed particles of dirt, so fine that they could not be seen visually. Dr Eric Jumper remarked that nothing could be more logical than to find dirt on the foot of a man who has walked without shoes[12].

Other tests showed that the bloodstains, unlike the image itself, did show capillarity, as would be expected from a liquid such as blood[13].

Studies were also made of the effects of the fire of 1532. Not surprisingly, direct observation of the Shroud showed a number of scorches from the fire. Studies of how these intersected with the image area indicated that neither the colour nor the density of the image changed as a result of being heated by the fire. There are few organic dyes or stains, or inorganic pigments that would have been available before 1532, that would have resisted the temperatures generated by the fire.

The water that was used to extinguish the fire was found to have migrated through the cloth in both scorched and unscorched image areas. Any medium used in paints, such as oils or waxes, would have affected this migration, another indication that no such media are present in the image[14]. In the end, the primary conclusion of the team was that "the image does not reside in an applied pigment"[15]. In other words, the Shroud is not a painting.

Other hypotheses of image formation were also addressed during the examination. The first was the Vapourgraph Theory proposed by Paul Vignon. The observations of the team ruled this out as a possibility. Vignon's theory was based on the image being formed by chemical transformation of foreign materials on the cloth. However, none of the physical or chemical properties of the materials suggested by Vignon (for example urea, ammonia or fatty acids) were observed on the Shroud. In addition, the diffusion transfer mechanism suggested by Vignon was found to be inadequate to explain the superficial nature of the image or to preserve the observed resolution of the image[16].

The scorch theory proposed by the British writer Geoffrey Ashe in 1966 was also investigated. This theory had already been discussed at the Albuquerque conference the previous year. Now the image on the Shroud was found by the team to have many of the properties of a light scorch. Scorches do not fade with time and are stable to further heating up to temperatures that will produce equivalent scorches in the base material. They do not move as water flows through them nor do they provide any barrier to water flow. In lightly scorched areas the fibres of a material are translucent, closely resembling the observed colour of the image-area fibrils on the Shroud. Some supporting evidence for this hypothesis was found by spectrophotometric work carried out by two of the team members, Roger and Marty Gilbert of Oriel Corporation, although there was disagreement on this from Sam Pellicori of the Santa Barbara Research Centre, another member of the team. The general conclusion of the team was that "the most likely scorch hypothesis is that the Shroud image is a light "air" scorch produced at temperatures lower than those sufficient to carbonize the material.... (However) any complete hypothesis must also account consistently for the observed density shading characteristics of the image."

Possible scorching techniques referred to include the hypothetical use of a "hot statue", based on the known 14th century existence of full-sized statues in either stone or metal. Theoretical studies carried out by Dr John Jackson had however shown that such a hypothesis is unlikely. Before the 1978 tests he had demonstrated mathematically that simple radiation emitted by a three–dimensional object could not have produced the density shading and resolution observed in the Shroud image[17]. Another possibility considered was the use of an etched or scribed flat plate, but the

Mark Oxley

team concluded that they were unaware of any "scorching technique that satisfactorily accounts for the observed image density characteristics."[18]

In addition to the studies made during the examination, samples were taken from the Shroud for further chemical investigation. These were taken from every feature, including off-image areas, various patches and burns, water stains, image and blood areas. They were lifted using a special tape, one side of which contained a chemically inert adhesive[19].

Finally, after 120 hours of sustained effort, the examination was finished on Friday 13 October. The STURP team packed up their equipment and returned home, with further work and studies to be done that would last for several years.

* * * * * * * * *

Dr John Heller was a biophysicist at the New England Institute in Connecticut in 1977. In his work there, he and his colleagues had developed a sensitive, new method of determining very small amounts of blood. He had contacted Dr Jackson to enquire whether his new method would be of use to the embryonic STURP team and had then been co-opted on to the team[20]. He was not in the party that went to Turin but awaited their return with samples to which he could apply his method. However, before he received any of the samples from the blood areas, nearly all the samples taken were passed for study to a leading American microscopist, Dr Walter McCrone.

Dr McCrone had made his name by a microscopic study he carried out in 1973 of the Vinland Map, a map showing America supposedly drawn a century before Christopher Columbus's voyage to the New World. He found that one of the substances in the ink of the map was a titanium-based chemical that had only been developed in 1920 and was thus able to declare the map a forgery. Dr McCrone had been introduced to the Holy Shroud Guild in 1974 and had tried unsuccessfully to obtain for examination some of the thread samples taken by Prof Gilbert Raes in 1973. He attended the Albuquerque conference in 1977 where he gave a presentation on recent developments in carbon-dating techniques and also claimed that with the use of extremely high resolution equipment, an ion microprobe, he might be able to identify from a fibre of the Shroud the nature of the

Shroud image and even the method of image formation[21]. He was signed on to the STURP team.

Unfortunately he then made a unilateral approach to ex-King Umberto II, seeking approval for his own carbon-dating initiative. As a result he was dropped from the main STURP team although his expertise was still to be used after the work in Turin was completed.

Dr McCrone started his examination of the Shroud samples at the end of 1978 and rapidly observed what he claimed to be evidence of the Shroud, after all, having been painted by an artist in a conventional manner. Examining the fibres taken from body-image areas, he determined that they were coated with particles of very finely powdered iron oxide. This is a natural artist's pigment that has been used since the earliest cave paintings. He also claimed to have found traces of a dried paint medium which he concluded to have been a collagen tempera, such as might have been made from boiling up scraps of parchment. He also found among the materials removed from the Shroud's surface what he identified as mercuric sulphide (artists' vermilion), ultramarine, orpiment and madder, all of which led him to the conclusion that the Shroud was, after all, no more than a painting[22].

In March 1979 the STURP team held its first post-examination meeting in Santa Barbara, California. The first paper was presented by Roger Morris, a physicist from the Los Alamos National Scientific Laboratories. He had carried out X-ray fluorescence tests on the Shroud and was able to show from these the presence of iron spread uniformly over the Shroud except in the blood areas, where it was higher than elsewhere. This would be expected in the event of real bloodstains, due to the iron atoms in blood. There was no measurable amount of inorganic substances. A further presentation by Ray Rogers suggested that there were no organic and biological substances present either.

When it was Dr McCrone's turn to present his results he stated that the body images had been made by red iron oxide earth pigments. In his opinion the iron oxide had been applied by a finger and the image was therefore a finger painting. He also stated that the blood, too, was made of an iron oxide paint. This flew in the face of the evidence gathered by scientists in the team that had traveled to Turin and the observations made by them. The reaction of Dr Sam Pellicori, an optical physicist from Santa Barbara Research Centre, who had measured the spectrum of iron oxide on numerous occasions, was that the colour from the image was totally wrong for what Dr McCrone was claiming. However, in the face of questions, Dr McCrone insisted that the X-ray fluorescence studies must be wrong and that purely by microscopic examination, without any specific chemical tests for iron oxide, he could confirm the Shroud as being a painting[23].

A second STURP meeting was held in late 1979, at Los Alamos. On this occasion Dr McCrone stated that the iron oxide was in extremely fine particles, most of them being less than one micron in size. He claimed that such fine particles of iron oxide had not existed prior to 1800 and that it was possible there may have been very faint images that some unknown person had "touched up" after 1800 with an iron oxide-gelatin paint to make them more visible[24].

A further meeting was scheduled at the Air Force Academy in Colorado Springs in March 1980. Dr Jumper told Dr McCrone that he would make available laboratories and any other equipment that would be required for Dr McCrone to enlarge on his findings. There would then be the opportunity for differences to be discussed and, if possible, resolved. Dr McCrone initially indicated that he would come. Now, he said, he had some new data. His initial finger painting theory was wrong and he had found the presence of a water solution of animal gelatin on his samples. However, he cancelled his appearance[25] and subsequently severed his connection with STURP. He eventually published his conclusions, together with a prediction that whenever the Shroud was carbon-dated the result would be a date around 1350. He never wavered in his opinions[26].

* * * * * * * * *

In turn Dr Heller studied samples taken from the Shroud. He had received four samples passed on to him by Dr McCrone. On initial microscopic examination of these samples he found traces of what he considered could possibly be blood, including a glob that he described as looking to him like "biltong" (sun-dried meat from Africa). He sought the assistance of a colleague, Prof Alan Adler, a physical chemist from Western Connecticut State University. By the time of the Santa Barbara meeting the two scientists had not yet been able to reach any firm conclusions.

Following the meeting Dr Heller carried out a further examination of his samples using a microspectrophotometer at Yale University. The spectrum of the "biltong" gave a reading that clearly indicated the presence of heme porphyrin, a constituent of blood. Dr Heller and Prof Adler reached the conclusion that there was hemoglobin on the samples, that it was denatured and very old[27].

Dr Heller and Prof Adler were able to do more chemical work at the Air Force Academy meeting in early 1980. The remainder of the samples taken from the Shroud in Turin had been collected from Dr McCrone and were now available for further study. Using sensitive chemical tests they were unable to find any trace of gelatins in the samples. They also carried out further tests on samples from the blood areas of the Shroud and satisfied themselves that the red particles on the fibrils were indeed blood. They later carried out yet more tests, using procedures that would be acceptable in a court of law for proof of the presence of blood. All were positive and the two men were able to go as far as determining that the blood was primate blood, most probably human[28].

By the time they completed their investigation, Dr Heller and Prof Adler had carried out over 1 000 experiments. They had succeeded in eliminating all paints, pigments, dyes and stains. They had been able to duplicate by chemical means the straw-yellow colour of the image, using what they described as a selective dehydrative acid-oxidising agent, but were unable to explain how the image could have been formed through such a process. They postulated that the image came from oxidation, but were unable to explain the chemical processes that could have led to such oxidation. They also ruled out heat as a source of the oxidation, as no heat source could account for the resolution or the three-dimensionality of the image, or the precision of the colour on the oxidized fibrils. They could not answer the question, "How did the image get there?"

They presented their findings at the annual meeting in 1981 of the Canadian Society of Forensic Sciences. Again Dr McCrone had been invited but declined to attend. He sent an assistant in his place. Prof Adler, in response to a question, could only agree that his and Dr Heller's data were at variance with Dr McCrone's[29].

Mark Oxley

* * * * * * * * *

At the same time as Dr Heller and Prof Adler were carrying out their studies on the blood samples, other scientific work was being done by other members of the STURP team. In addition, explanations were being sought for Dr McCrone's reported findings.

Other ways in which the linen fibres could have been coloured without the use of pigments were investigated. Dr Sam Pellicori hypothesized that a catalyst had sensitized the Shroud to produce a latent image. This had then been developed at a later stage by heat or by the aging of the linen, to produce the visible image. He simulated the aging process by baking linen samples in an oven, applying thin coatings of natural products including skin secretions and then putting them back in the oven. He found that the treated areas did develop a yellower colour than the untreated areas. This was one possible way in which the Shroud could have been artificially created. However this hypothesis could not explain the three-dimensional nature of the image. Another member of the STURP team, John German of the US Air Force Weapons Laboratories, proposed a variation to Dr Pellicori's hypothesis to try and overcome this problem, but his variation was still unable to explain the actual differences in image density on the Shroud[30].

Dr Ray Rogers and Dr Larry Schwalbe of Los Alamos National Scientific Laboratories had carried out their own microscopic examination of the fibrils. From the observed fibril structure they concluded that there was no obvious evidence for a coating of a paint medium. If a very dilute or non-viscous medium was used, then there should have been discolouration deeper into the threads, as well as liquid capillary flow. Neither was observed. As a result of their work they were quite naturally sceptical about the results reported by Dr McCrone.

In reviewing the scientific work done, Dr Rogers and Dr Schwalbe noted that chemical tests showing evidence of proteins in the blood areas, as carried out by Dr McCrone and by Dr Heller and Prof Adler, could indicate the presence of either a protein-based paint medium or blood. However Dr McCrone had failed to show the presence of any proteins in pure-image areas where there was no blood. They also noted that the chemical tests used by Dr McCrone could easily yield false positive results[31].

Other reasons were advanced for the presence of iron oxide particles on the Shroud. Iron oxide could have been a contaminant from contact with the protective glass plates used in earlier exhibitions of the Shroud. It could also have been produced during the original linen retting process when the Shroud was first made – this hypothesis has been supported by detection of iron and calcium bound by ion exchange on both image and non-image fibrils. Iron oxide could also have been produced at the time of the 1532 fire, when the Shroud was doused with water. Iron oxide-coated fibrils seem to occur most frequently in the samples from the margins of the water stains.

There is also difficulty in distinguishing optically between iron oxide particles and "blood sherds" or "blood flakes" as reported by Dr Heller and Prof Adler. Dr McCrone reported large concentrations of iron oxide particles in the blood areas, determined by microscopic optical methods. Dr Heller and Prof Adler found most of these particles, through chemical testing, to be "blood flakes".

Dr Heller and Prof Adler noted that iron oxide found in paint-type pigments is invariably contaminated with manganese, nickel, cobalt or aluminium. Tests they carried out on the particles in the Shroud samples revealed no significant traces of these elements. They concluded that the iron oxide came from either the water used at the time of the 1532 fire or was of biological origin. Some traces of silver that they found could be ascribed to molten debris from the casket falling on to the Shroud during the 1532 fire.

With regard to other paint-type materials reported by Dr McCrone, Dr Heller and Prof Adler noted that the occasional occurrence of such materials on the Shroud should be expected, as the Shroud is known to have been copied by artists and was in all likelihood therefore exposed to the paints that they used[32].

* * * * * * * * *

The final meeting of the STURP team took place in New London, Connecticut, in late 1981. It marked the third anniversary of the work done in Turin and was the last occasion on which the team met as a unit.

Dr McCrone was invited to participate but once again he declined. The team expressed its hopes that a carbon-14 dating test would be carried out sometime in the future[33]. Following this meeting STURP issued its Final Report. The following were its official conclusions:

"No pigments, paints, dyes or stains have been found on the fibrils. X-ray, fluorescence and microchemistry on the fibrils preclude the possibility of paint being used as a method for creating the image. Ultra-violet and infrared evaluation confirm these studies. Computer image enhancement and analysis by a device known as a VP-8 image analyzer show that the image has unique three-dimensional information encoded in it. Microchemical evaluation has indicated no evidence of any spices, oils or any biochemicals known to be produced by the body in life or in death. It is clear that there has been a direct contact of the Shroud with a body, which explains certain features such as scourge marks, as well as the blood. However, while this type of contact might explain some of the features of the torso, it is totally incapable of explaining the image of the face with the high resolution that has been amply demonstrated by photography. The basic problem from a scientific point of view is that some explanations which might be tenable from a chemical point of view are precluded by physics. Contrariwise, certain physical explanations which may be attractive are completely precluded by the chemistry. For an adequate explanation for the image of the Shroud, one must have an explanation which is scientifically sound, from a physical, chemical, biological and medical viewpoint. At the present, this type of solution does not appear to be obtainable by the best efforts of the members of the Shroud Team. Furthermore, experiments in physics and chemistry with old linen have failed to produce adequately the phenomenon presented by the Shroud of Turin. The scientific consensus is that the image was produced by something which resulted in oxidation, dehydration and conjugation of the polysaccharide structure of the microfibrils of the linen itself. Such changes can be duplicated in the laboratory by certain chemical and physical processes. A similar type of change in linen can be obtained by sulphuric acid or heat. However, there are no chemical or physical methods known which can account for the totality of the image, nor can any combination of physical, chemical, biological or medical circumstances explain the image adequately.

"Thus, the answer to the question of how the image was produced or what produced the image remains, now, as it has in the past, a mystery.

"We can conclude for now that the Shroud image is that of a real human form of a scourged, crucified man. It is not the product of an artist. The blood stains are composed of hemoglobin and also give a positive test for serum albumin. The image is an on-going mystery and until further chemical studies are made, perhaps by this group of scientists, or perhaps by some scientists in the future, the problem remains unsolved."[34]

Thus three years of work by over thirty reputable scientists came to an end. More than twenty scientific papers on their work on the Shroud were published by members of the team in peer-reviewed publications in the period 1978 – 1984. The STURP team was not, as some sceptics have suggested, a group of dedicated Christians committed to proving the authenticity of the Shroud. They were scientists drawn from a range of reputable scientific and academic institutions. Their personal beliefs ranged from Catholic to Protestant and from agnostic to Jewish. At the start of the project many of them had believed that they would prove quite easily that the Shroud was no more than some sort of mediaeval painting. At the end, they had shown clearly that it was not, but had not been able to determine exactly what it was; it was a mystery to the science of the day.

STURP itself was a non-profit organization supported solely by contributions from private individuals. The research carried out was the product of volunteer efforts by all the project members.

And what about Dr McCrone? The Final Report of STURP totally rejected his findings and reached conclusions diametrically opposed to his. If the conclusions of STURP in their totality are valid, then he must be wrong. How could a distinguished scientist of unquestioned integrity get it so wrong (assuming that he did)? The answer to this question probably lies again in the challenge posed by the Shroud. The idea that the Shroud might be the authentic burial cloth of Jesus Christ has immense consequences for one's religious beliefs, with its implications of the Resurrection. For those who are agnostics, or merely comfortable with not having to think too much about spiritual or metaphysical matters, this is an unacceptable idea. Scientists, like everyone else, can have a tendency to see what they want to see and to overlook what might be too radical or frightening. In the interests of science they shouldn't, but sometimes they do. Perhaps Dr McCrone was simply unable to accept the intellectual and spiritual consequences of finding that the Shroud of Turin might – just might – be authentic[35].

Chapter 16 – Dating the Shroud

On 13 October 1988 an announcement was made in the Press Room of the British Museum in London that shocked the many believers in the authenticity of the Shroud and seemed to bear out the scepticism of those who saw it as a 14th century fake. Dr Michael Tite of the British Museum's Research Laboratory, together with Professor Edward Hall of the Research Laboratory for Archaeology and History of Art at Oxford University and Dr Robert Hedges, an associate of Professor Hall, announced that radiocarbon dating tests that they had carried out on the Shroud showed it to date from the period 1260 – 1390, and that the Shroud's raw flax had probably been made into linen in about the year 1325, with a small margin of error.

If this was so, clearly the Shroud could not be the genuine shroud of Jesus Christ, but was the product of an ingenious and inventive fourteenth-century forger. There could be no doubt that the scientists involved in these tests, as well as their laboratories, were of the highest integrity and international reputation. The scientists themselves had no doubts about the validity of their results. Professor Hall himself remarked, "There was a multi-million pound business in making forgeries during the fourteenth century. Someone just got a bit of linen, faked it up and flogged it."[1] He also described anyone continuing to regard the Shroud as genuine as a "Flat Earther" and "on to a loser".[2]

To the scientists involved and the sceptics, who included much of the Press and many subsequent writers on the Shroud, this was the end of the matter. To others, however, there remained some unanswered questions. The first was this: the evidence gathered by the STURP team during their study of the Shroud seemed to indicate that it was the genuine shroud of a man who had been crucified in a manner consistent with the death of Jesus Christ. If 95% of the evidence pointed in one direction and only 5% - the radiocarbon dating – pointed in another, which should be given credence? The second question was, if the Shroud was indeed a forgery, how was it forged? Nobody has yet been able to provide a satisfactory answer to this question, despite numerous theories being advanced – some ingenious, some fanciful in the extreme. These led in turn to questions about the reliability of the radiocarbon dating tests and various theories about why the tests had to have yielded an incorrect result.

The scientific theory on which radiocarbon dating is based has been explained in a sufficient number of books and papers. There is no need to describe it yet again. It is a scientifically sound theory which has yielded many accurate results that have greatly advanced knowledge in fields such as archaeology and history.

* * * * * * * * *

The technology used for the tests on the Shroud, accelerator mass spectrometry (AMS), had first been successfully tested for carbon dating in 1977 by Professor Harry Gove of Rochester University in the United States. Before this development, carbon dating required relatively large samples of material. The new technique allowed tests to be carried out on very small samples. Carbon dating was not used by the STURP team in 1978 as it would have required the destruction of too large a sample. Professor Gove's method was too new at the time to have been considered.

However it was not long before the idea arose of using this method to definitively answer one of the two great questions about the Shroud – how old is it? Professor Gove discussed the possibility with Professor Hall in Oxford in late 1978. A few months later he and his colleagues wrote to Cardinal Ballestrero, the Archbishop of Turin, formally offering to use their new method to establish the age of the Shroud[3].

Matters progressed slowly. In 1983, the feasibility of dating the Shroud using the AMS method was confirmed by a successful intercomparison, involving four AMS laboratories and two laboratories using the small gas-counter method and the dating of three textile samples of known age, co-ordinated by the British Museum[4].

By 1986 a set of recommendations for the radiocarbon dating of the Shroud were set up, known as the 1986 Protocol. This recommended that seven radiocarbon-dating laboratories, of which five were AMS and two small gas-counter, which had expressed interest in carrying out the dating and which had been involved in the preparation of the Protocol, should all be given Shroud samples of a suitable size for testing. It was further recommended that a Swiss textile expert, Mme Mechthild Flury-Lemberg of the Abegg Foundation in Bern, should perform the sample removal. The samples would be taken from sites away from an area charred by fire and also away from any area of obvious scientific information value. Dr Tite of the British Museum's Research Laboratory was to supervise the process. He would also provide "control" samples of a known age and similar in texture to the Shroud so that the testing laboratories would not know which of the samples came from the Shroud. This would ensure that the testing would be scientifically "blind".

However, matters were not to be so simple. Professional jealousies and disputes within the Church raised their heads. In October 1987 Cardinal Ballestrero overruled the Protocol and decided that only three laboratories would participate – those in Oxford, Arizona and Zurich – and that Mme Flury-Lemberg would not be involved in the sample removal. One of the rejected laboratories was that of Professor Gove, who was sufficiently aggrieved by his rejection that he sought, unsuccessfully, to raise the matter with Pope John Paul II[5].

The samples were taken from the Shroud at Turin Cathedral on 21 April 1988. Cardinal Ballestrero was present, together with two textile experts, representatives of the three selected laboratories and Dr Tite, who had been chosen to oversee the testing. The sample was removed from the Shroud by Prof Giovanni Riggi, a microanalyst from Turin. A strip was cut from just above the place where Professor Raes had taken one of his samples in 1973. It came from a single site, away from any patches or charred areas, and three samples were prepared from this strip[6].

Three control samples were provided – one from the eleventh to twelfth centuries, from Egypt; one from about the first century AD, also from Egypt; and one from the cope of St Louis d'Anjou in France, dated to the beginning of the fourteenth century. The Shroud samples were examined microscopically to identify and remove any contaminants and then cleaned in various ways at the different laboratories.

Testing was carried out at each laboratory of the three control samples and of the sample of the Shroud. The three control samples were all dated correctly. The results of the testing on the Shroud samples yielded a calibrated calendar age range with at least 95% confidence for the linen of the Shroud of 1260 – 1390. The scientists concluded that:

"These results therefore provide conclusive evidence that the linen of the Shroud of Turin is mediaeval."[7]

* * * * * * * * *

Even before the radiocarbon tests were carried out, some doubts had been expressed about the possibility of obtaining a valid result from such tests. Prof William Meacham, an archaeologist at the University of Hong Kong, expressed grave concern in 1986 about the possibility that contamination in any samples would cause major errors in the dating results. He warned that the original proposal by Professor Gove for carbon dating the Shroud outlined only standard pre-treatment of the samples. It did not propose scanning electron microscope screening or other types of direct examination to check the state of the samples prior to testing[8]. All three testing laboratories did examine their samples microscopically before carrying out mechanical and chemical cleaning procedures, but the microscopical examinations made were not necessarily as accurate as the scanning electron microscope screening proposed by Dr Meacham. The possibility of the samples having retained contaminants that were not removed by the testing laboratories cannot be ruled out.

Contaminants made up of porous materials in particular are difficult to remove. There have been numerous widely-cited examples of incorrect radiocarbon dating which may well have been caused by such contaminants. In 1985 the Swiss laboratory in Bern that was later selected to be one of the testing laboratories for the Shroud had itself made a serious error, resulting in a misdating of 1 000 years[9]. Another example is that of Egyptian Mummy no 1770 in the collection of the Manchester Museum. This mummy was unwrapped in the 1970s and its body tissues and bandages sent to the British Museum's laboratory for carbon dating. There was a discrepancy of 1 000 years between the apparent ages of the body and the bandages[10]. A similar instance occurred in 1996, when a mummy of an ibis was similarly dated. In this case there was an apparent discrepancy of 550 years between the age of the bird and the age of its wrappings. Professor Gove himself took part in the tests on the ibis[11]. Despite attempts to find alternative explanations for these differences, the most likely conclusion is that the differences were caused by some undetected contaminants in the linen wrappings of the mummies.

A specific theory on possible contamination was developed by an American microbiologist, Dr Leoncio Garza-Valdes. In studies of pre-Colombian American artifacts he found a coating formed from bacteria and fungi on their surfaces. These bacteria and fungi formed a yellowish "plastic coat", which he described as a "bioplastic coating" in view of its bacterial origin. He found that the coating was a polyester known as polyhydroxyalkanoate, or PHA. A radiocarbon dating of one of the artifacts produced a result that he found questionable and he put this discrepancy down to the effect of the bioplastic coating[12].

Dr Garza-Valdes then questioned whether the date for the Shroud established by the radiocarbon testing may also have been affected by a similar contamination. He was able to obtain from Prof Riggi in Turin, who had cut the samples for testing, samples of the trimming from the edge of the Shroud that he had retained at the time of cutting. He had kept these because he had thought at the time that the edge of the Shroud might be contaminated. On examining these samples with an optical microscope, Dr Garza-Valdes saw what he described as a bioplastic coating on the fibres. Prof Riggi confirmed this observation[13]. The two scientists subsequently worked together on removing the coating from the fibres. Dr Garza-Valdes claimed that the plastic coating made up more than 60% of the fibre material. He found this highly significant, as Professor Hall, in defending the validity of the radiocarbon dating of the Shroud, had stated that a contamination of more than 60% would be needed to invalidate the radiocarbon dating results. Dr Garza-Valdes concluded that the 1988 samples used in the dating of the Shroud had cellulose from the flax as well as bacteria, a small number of fungi and the bioplastic coating[14]. The presence of such contaminants was, in his view, sufficient to distort the results of the radiocarbon dating.

Dr Garza-Valdes and Prof Riggi's final step was to prepare samples of the Shroud fibre without the contaminants, for testing. Samples were given to the AMS laboratory at the University of Arizona, which was one of the laboratories that had carried out the original dating of the Shroud sample, and to a similar facility at the University of Toronto. The results were available in December 1994. The University of Arizona had dated the sample as approximately 5 000 years old – 3 000 years before the birth of Jesus – and the University of Toronto had dated it some 800 years younger, but still too old for the Shroud. Dr Garza-Valdes has since conceded that his sample had in its turn been accidentally contaminated by one of the chemicals used in the preparation of the samples. However, he has also stressed the point that the samples tested by the two laboratories were identical and he has expressed his view that the difference in their results by itself casts doubt on the validity of radiocarbon dating of such textile samples[15].

* * * * * * * * *

Further doubt was cast on the radiocarbon dating results in a paper presented in August 2000 at an international congress entitled "Sindone 2000" held in Orvieto, Italy. The paper was presented by Joseph G Marino and Sue Benford and suggested that the sample taken from the Shroud for the radiocarbon testing had contained 16th century material spliced into the Shroud cloth for the purpose of repairs. This had therefore altered the overall date of the sample to make it appear more modern than the original Shroud material.

Specifically, the authors of the paper suggested that a section of the Shroud was removed in 1531 as a result of the execution of the last will and testament of Margaret of Austria, Duchess of Savoy, referred to in Chapter 7. The Duchess Margaret stipulated in her will of 1508 that a portion of the Shroud be given to her church of Brou at Bourg-en-Bresse. Whether this bequest was actually executed is not known[16], but it is the contention of the authors of the paper that it was carried out. They suggest that a section measuring five and a half inches by three and a half inches was removed and replaced by a patch carefully woven into the material. Further, this section is adjacent to the sample taken for testing in 1988.

In support of this hypothesis they described in detail anomalies in the cloth in the area from which the sample was taken, including more pronounced discolouration in the area and the presence of starch. Starch was used by mediaeval restorers for invisible mending. They also showed the existence of a vertical seam in the vicinity of the sample, which could indicate the connecting point of the patch to the Shroud, as well as weave pattern inconsistencies in the area.

Finally, the authors made the point that a merging of threads from AD 1500 into a 2 000 year old piece of linen would augment the carbon-14 content, such that a 60/40 ratio of new material to old, determined by mass, would result in a carbon-14 age of approximately AD 1210. This correlates very closely with the mean date of AD 1200 determined by the Oxford laboratory[17] and with the ratio of original versus mediaeval material in the carbon-14 sample area, according to the studies reported by the authors[18].

In a further paper in 2002, the authors looked further at the historical evidence for the removal of a portion of the original Shroud and its replacement by a mediaeval patch. They referred to the sophistication of the weaving art and the tapestry industry of the 16th century. Duchess Margaret was a patron of the tapestry industry and her nephew and ward, the Emperor Charles V, also took a great interest in tapestry as an art. The authors suggested that Charles may have been directly involved in the execution of her bequest of a portion of the Shroud and in the subsequent replacement of that portion with a patch. Charles would certainly have had the resources and skills at his disposal to ensure that the Shroud was repaired to its original quality.

A Shroud replica commissioned in 1571 by Pope Pius V for Don Juan of Austria, the illegitimate son of Charles V and great-nephew of the Duchess Margaret, shows a corner piece as being missing. It can be speculated that this may be seen as the provision of evidence to Don Juan that his great-aunt's bequest had been carried out.

The authors also quoted an expert on mediaeval weavers and tapestries, Robert Buden, the President of Tapestries and Treasures in the United States of America, as stating that weavers of the 16th century definitely possessed the skill to carry out invisible repairs of textiles, and that repairs to a relic such as the Shroud would undoubtedly have been assigned to the most skilled craftsmen, with no expense spared[19].

These theories inevitably met with criticism. The authors published a third paper responding to these criticisms. In particular they referred to work done by the former STURP chemist Dr Ray Rogers of the Los Alamos National Laboratory in New Mexico, who had samples of both the threads removed from the Shroud by Professor Raes in 1973 and of threads removed from the main part of the Shroud during the STURP investigations in 1978. Dr Rogers carried out detailed chemical studies on the samples in his possession and concluded that:

"The combined evidence from chemistry, cotton content, technology, photography and residual lignin proves that the main part of the Shroud is significantly different from the radiocarbon sampling area. *The validity of the radiocarbon sample must be questioned with regard to dating the production of the main part of the cloth.*"[20] (Author's italics.)

In other comments Dr Rogers noted again that the samples used for the dating of 1988 were cut from immediately above the Raes sample and remarked that the evidence shows that it is highly probably that the samples (used in the radiocarbon dating) were not characteristic of the Shroud and were spurious[21].

In their third paper Benford and Marino also publicized an unauthorized radiocarbon dating test that had been carried out in 1982 using a single thread from the Raes sample. The dating was carried out by Dr George Rossman, a mineralogist at the California Institute of Technlogy. The thread was supplied by Prof Alan Adler, who was apparently unaware at the time that an agreement had been signed by STURP members not to do further testing on Shroud samples. Prof Adler informed Dr Rossman that one end of the thread contained what he described as appearing to be a "starch contaminate". Dr Rossman cut the thread in half and dated each end of the thread separately. He found that the non-contaminated end of the thread dated to 200 AD and the starched end to 1200 AD. These data were not published at the time but Dr Rossman did confirm to Sue Benford in a personal communication that he had carried out these tests[22].

In microscopic examinations of a Raes thread Dr Rogers observed what appeared to be a splice in the middle of the thread. One end of the thread showed encrustation and colour while the other end was nearly white. He described this as being "obviously an end-to-end splice of two different batches of yarn" and noted that no splices of this type were observed in the main part of the Shroud[23].

Other microscopic examinations of Raes threads were carried out by Dr John L Brown, a retired scientist who had worked at the Georgia Technical Research Institute in Atlanta, Georgia. He noted a yellow-brown coating that he felt was a clear attempt by a mediaeval artisan to dye a newly added repair region of fabric, to match the aged appearance of the remainder of the Shroud. He also noted cotton fibres, which had also been noted by other researchers[24]. This is significant, as cotton was virtually unknown in Europe until at least the middle of the 14th century and would therefore have been most unlikely to be present in a cloth made in Europe in the late 13th or early 14th century.

These observations and tests would appear to support the hypothesis of Benford and Marino that a patch was indeed inserted in the Shroud, in the area of the Raes threads and the radiocarbon dating sample.

Benford and Marino certainly had no intention of letting this subject drop. They produced yet another paper in 2005 detailing further evidence of an invisible patch in the radiocarbon sample area of the Shroud and giving further responses to criticisms of their hypothesis. One particular critic, at the Third International Dallas Shroud Conference in September 2005, was the textile expert who had overseen the 2002 restoration of the Shroud, Mme Mechthild Flury-Lemburg. She stated categorically that there is no reweave in the Shroud. She had previously told the authors that the invisible repair they were suggesting was "technically impossible". In response to this the authors contacted an expert on modern invisible mending, Michael Ehrlich of Chicago. Mr Ehrlich made a distinction between "Inweaving", which is the modern technique to which Mme Flury-Lemburg appeared to have been referring, and "French Weaving", a technique used in 16th century Europe which does result in "invisibility" of the repair[25].

Mme Flury-Lemburg has continued to maintain that no such repairs have ever been made to the Shroud. In a paper published in 2007, she concluded:

"....neither on the front nor on the back of the whole cloth is the slightest hint of a mending operation, a patch or some kind of reinforcing darning, to be found."[26]

Mme Flury-Lemburg's conclusion is clearly at odds with the observations of Dr Ray Rogers referred to above, who has reported a clear splice in one of the Raes threads. She has, however, been supported in her views by the President of the Diocesan Commission for the Holy Shroud, Mgr Giuseppe Ghiberti. On the other hand the point has also been emphasized that any sixteenth century invisible re-weave would not be visible to even a close examination by the naked eye and would only be discoverable through microscopic and microchemical analysis, such as that carried out by Dr Rogers[27].

An examination carried out by the STURP team in 1978 using a technique known as Quad-Mosaic Photography has also been shown to yield evidence of differing chemical compositions of the material in the area where the samples were taken for the radiocarbon testing[28].

Benford and Marino have also quoted the scientific advisor to the Archbishop of Turin, Professor Piero Savarino, as having written in a booklet some years previously that the 1988 radiocarbon dating may have been erroneous due to:

"....extraneous thread left over from 'invisible mending' routinely carried out in the past on parts of the cloth in poor repair.... If the sample taken had been the subject of 'invisible mending' the carbon-dating results would not be reliable. What is more, the site from which the samples actually were taken does not preclude this hypothesis."[29]

* * * * * * * * *

Margaret of Austria was the daughter of the Holy Roman Emperor Maximilian I. In 1501 she married Duke Philibert II of Savoy. The marriage was short-lived, as Philibert died in 1504. As the Dowager Duchess,

Margaret became the custodian of the Savoy collection of tapestries, which was extensive, as well as of the relic now known as the Shroud of Turin.

Margaret was known for her love for beautiful tapestries. One authority has noted that:

"Margaret... exercised an important influence over the development of the Netherlands tapestry industry during a crucial phase in its development. Although she lacked the funds that her ancestors enjoyed, Margaret evidently shared their love of the tapestry medium, ensuring the careful preservation of the inherited Burgundian collection during her reign (as Regent of the Netherlands, prior to her marriage to the Duke of Savoy) and developing a small but interesting collection of her own."[30]

In 1508 Margaret drew up a will in which she stated:

"I give to my church St Nicholas (in Brou) all the holy relics I have now and will have on the day of my death, the piece of the Holy Cross, the Holy Shroud, bones of saints that I have and will have on my death, and which could decorate the church."[31]

The building of a church in Brou and rebuilding of the ruined Benedictine monastery there had been the deathbed request of Margaret's husband, Duke Philibert. This request in turn stemmed from a commitment and promise made by his mother, Margaret of Bourbon, which had never been carried out. This commitment became almost an obsession with Margaret during the remainder of her life. Her dedication was not, however, shared by her counselors or family, who were deeply concerned about the expenses involved. One of these counselors was Laurent de Gorrevod, the Governor of Bresse. He, and other counselors, tried to persuade Margaret that her money would be better spent on restoring the already-begun Notre Dame church in Bourg. Margaret refused to change her mind.

In 1530 Laurent de Gorrevod's brother Louis accepted a position at the Notre Dame church as a newly-appointed cardinal. When Margaret died later that year he was in charge of all matters related to the Catholic Church in the region. Margaret's will stated that the entire Shroud (rather than just a portion, as suggested by some writers) should be given to the church at Brou. However, neither her Savoy family nor her nephew the Emperor Charles V had any interest in this church, which was still unfinished, and it may not be unreasonable to hypothesise that an agreement was reached, possibly between Charles V and Cardinal de Gorrevod, to cut a small piece from the Shroud to give to the church in Brou, while the Shroud itself remained in the hands of the Savoy family.

If this was done, the repairs to the Shroud where the piece was taken would undoubtedly have been entrusted to the most skilled tapestry weavers of the time, whom Charles V certainly had access to. The best-known tapestry-worker of the period was Pieter van Aelst, whose tapestry workshop had been contracted to produce nine tapestries used at Charles's coronation. Again, it is not unreasonable to suggest that van Aelst may have been entrusted with the necessary repairs to the Shroud[32].

This is of course all historical speculation and surmise. There is no direct evidence to support this hypothesis. There is, however, direct scientific evidence that repairs have been made to the Shroud in the area where the sample was taken for the radiocarbon dating tests. This small piece of history may provide a plausible explanation for this.

* * * * * * * * *

With the evidence suggesting that the radiocarbon dating tests used at least some mediaeval material patched into the Shroud, what of the theory of Dr Garza-Valdes that a "bioplastic" coating on the Shroud material could have been responsible for an erroneous dating? This was taken up by Dr Rogers and Anna Arnoldi of the University of Milan in a paper which reviewed how the Scientific Method had been applied in studies of the Shroud. In this paper the authors remarked that:

"Failure to test against all observations can lead to a logically consistent but unsupportable superstructure."[33]

The authors criticized Dr Garza-Valdes' theory on two major grounds: firstly, any organic carbon added to the Shroud, resulting in a significant error in the radiocarbon dating, must have been rich in carbon-14 and hence fairly modern. In addition the amount of modern carbon added must be large. Secondly, chemical analysis of the Shroud material and of the coating itself shows that the chemical conditions for the formation of a bioplastic coating do not exist in the Shroud. In particular they pointed out that the coating on thread samples from the Shroud examined by the STURP team contained a chemical called alizarin and emphasised that any gum coating containing alizarin cannot be a "bioplastic polymer".

They also made the point that the Shroud has always been stored dry and usually out of direct light in some kind of container. This would have minimized the possibilities for the formation in the Shroud cloth of organisms that could absorb carbon from the atmosphere. Dr Rogers also measured the thickness of the coating on thread samples from the Shroud. The thickness does vary, but he estimated that overall it would contribute only a few percent to the weight of the Shroud – certainly not enough to produce a significant error in the age determination.

Dr Rogers suggested that the gum-dye coating found in the area where all the thread samples were taken had been mistaken by Dr Garza-Valdes for a "bioplastic coating" and ended by remarking that:

"Unfortunately, the hypothesis was developed without a rigorous application of Scientific Method."[34]

* * * * * * * * *

In January 2005 Dr Rogers published a paper in the peer-reviewed scientific journal *Thermochimica Acta* in which he described studies he had carried out on threads from the samples taken for the radiocarbon dating. He received these samples from Prof Luigi Gonella of Turin Polytechnic University, who in turn had taken them from the centre of the radiocarbon sample before it was distributed for testing. He also had thread samples from other areas of the Shroud which he had taken himself in 1978 when he was a member of the STURP team. He was thus in a position to compare the chemical composition of the threads from the radiocarbon sample area with threads from other areas of the Shroud.

The flax fibres in the Shroud material contain a substance called lignin, which in its turn contains a chemical called vanillin. Over an extended period of time the lignin loses its vanillin and the rate of this loss can be used to estimate an age for any material containing lignin. Dr Rogers found that vanillin cannot be detected in the lignin on the Shroud fibres that he took in 1978. Based on the known rate of vanillin loss, he concluded that this means that the Shroud is at least 1 300 years old and a maximum of 3 000 years old.

Tests that he carried out on mediaeval linens, as well as on thread samples taken in 1973 by Prof Gilbert Raes from the same area of the Shroud as the radiocarbon samples, clearly showed the presence of vanillin. This suggested to him that the Raes threads contain material from mediaeval times.

He also showed that the Raes and radiocarbon sample threads are coated with a plant gum containing alizarin dye as well as colloidal red dye lakes. The presence of this dye and the red lakes indicate that the colour of the threads has been manipulated. In other words the implication is that repairs were made at some unknown time with foreign linen dyed to match the older original material. This is totally consistent with the hypothesis of Benford and Marino described earlier in this Chapter.

Dr Rogers' conclusion was summed up in the following terms:

"Preliminary estimates of the kinetics constants for the loss of vanillin from lignin indicate a much older age for the cloth than the radiocarbon analyses. The radiocarbon sampling area is uniquely coated with a yellow-brown plant gum containing dye lakes. Pyrolysis-mass-spectrometry results from the sample area coupled with microscopic and microchemical observations prove that the radiocarbon sample was not part of the original cloth of the Shroud of Turin. *The radiocarbon date was thus not valid for determining the true age of the Shroud.* (Author's italics)"[35]

Dr Rogers' conclusions were reported on the BBC News on 26 January 2005 and then promptly forgotten by the world at large. Today there remains a general belief that the radiocarbon dating of 1988 proved beyond doubt that the Shroud is a mediaeval fake.

In fact this whole story is not just one of complacent misbelief; it also includes examples of scientific arrogance and narrow-mindedness.

The first major defect of the radiocarbon testing was that only one sample was taken from the Shroud. This sample was then divided into three and the three pieces were sent to different laboratories for testing. Whatever the reasons behind the taking of a single sample, it was poor science. The normal scientific procedure would be for several samples to be taken from different parts of the object being tested, so that any contamination or defect in one sample could easily be detected through an obviously false result. In this case the single sample used was contaminated but the result was wrongly accepted as being accurate.

The second major defect arose from the fact that scientists can be very narrow-minded about their own specialist fields of study. They sometimes have a tendency to believe that only their speciality can provide accurate results and that other fields of study are of less importance or validity. This was certainly the case with the scientists responsible for the radiocarbon dating.

The examination of the Shroud by the STURP team in 1978 had found clear evidence that the Shroud could possibly be a genuine first-century burial cloth. This evidence was totally overlooked and ignored by the radiocarbon dating scientists when they declared in absolute terms that:

"These results therefore provide conclusive evidence that the linen of the Shroud of Turin is mediaeval."

They provided no such "conclusive" evidence. Their results merely produced one conclusion – although a compelling one - in a series of studies on the origin, nature and age of the Shroud. Where the bulk of scientific evidence gives one indication, as was the case with the STURP studies, and another study gives a conflicting result, as did the radiocarbon testing, then all the evidence needs to be reviewed in its entirety and an explanation sought for such discrepancies. This was not done. Professor Hall's remark quoted at the beginning of the Chapter that anyone who still considered the Shroud to be genuine was a "Flat Earther" and "on to a loser" is a good example of scientific arrogance and narrow-mindedness. Needless to say, the work of Dr Rogers in particular has overtaken Professor Hall's views.

Dr Rogers' study of the loss of vanillin in the Shroud material has proved an ingenious alternative dating method for the Shroud, but it is by no means precise. Radicarbon dating is a precise method which will give an accurate age for the Shroud, provided that genuine samples of original Shroud material are tested under rigorous scientific conditions. One must hope that such testing will be carried out eventually, in order to resolve this controversy once and for all.

Chapter 17 –
Image Formation: Radiation Hypotheses

The second great question about the Shroud that science seeks to answer is: how was the image formed? Finding the answer to this question was one of the prime objectives of the Shroud of Turin Research Project in 1978. Despite detailed study by leading scientists in their respective fields, STURP concluded in its Final Report, as we have seen, that:

"…there are no chemical or physical methods known which can account for the totality of the image, nor can any combination of physical, chemical, biological or medical circumstances explain the image adequately. Thus, the answer to the question of how the image was produced or what produced the image remains, now, as it has in the past, a mystery."[1]

Over the years following the STURP report numerous theories and suggestions have been put forward by reputable scientists as to how the image was formed. A number of these scientists were members of the STURP team and have been following up on the work done by STURP. These theories can be divided into two basic classes – those which suggest that the image was formed by physical means, such as radiation, and those which suggest that the image was formed by chemical or biochemical means.

The STURP Final Report also made the following cautionary remarks:

"The basic problem from a scientific point of view is that some explanations (for the formation of the image) which might be tenable from a chemical point of view are precluded by physics. Contrariwise, certain physical explanations which may be attractive are completely precluded by the chemistry. For an adequate explanation for the image of the Shroud, one must have an explanation which is scientifically sound, from a physical, chemical, biological ad medical viewpoint."[2]

The only scientific consensus that has been reached so far on the question of image formation is that it is an extremely complex matter. Indeed, it may go beyond the bounds of science as we know it and may be even more complex than we can possibly understand.

* * * * * * * * *

STURP concluded that its studies precluded the possibility of paint being used as a method for creating the image. It is not a painting and it has unique three-dimensional information in it. The bloodstains on the Shroud are real human blood[3].

STURP also found that the image resulted from a yellowing of the topmost fibres of the threads of the cloth. Furthermore, the yellowing of the fibres penetrated into each thread by only the depth of one fibre. In other words the image is extremely superficial. The yellowing of the fibres is also uniform. The difference in the shading of the image is caused by an increased concentration of yellowed fibres in the darker areas, rather than by a difference in the degree of yellowing of the individual fibres[4].

Medical forensic examinations have been made of the blood images. These are all in agreement that they are bloodstains from clotted wounds transferred to the cloth through direct contact between a wounded human body and the cloth[5]. A medical consultant and researcher who has studied the Shroud for over twenty years, Dr Gil Lavoie, carried out studies on the transfer of clotted blood to cloth. Once blood clots are formed they shrink slightly in size, at the same time squeezing out a clear yellow fluid called serum. His experiments showed that the man on the Shroud must have been vertical when he bled and died. Further, it was the moisture of the serum that allowed the transfers to take place on to the cloth. These transfers could only take place up to a limited time after clotting, otherwise the clots became too dry to be transferred. This time limit, he estimated, would be between one and a half and two hours. It was his conclusion, therefore, that the bloodstains on the Shroud would have been made up to two hours after the death of the man on the Shroud[6].

Image Formation: Radiation Hypotheses

The studies of the STURP team had shown that when the blood was removed from blood-covered cloth fibres in the image area of the Shroud, the fibres underneath were white. This clearly showed that the blood went on to the cloth before the image marks and that the blood protected the fibres from whatever caused the image. The blood came first, then the image[7].

Dr Lavoie carried out further studies in which he demonstrated that the blood marks on the Shroud are consistent with a cloth having been draped and tucked over a man's face covered with moist blood. The transfer of the blood to the cloth was indeed a simple contact process[8]. However, he also found that in the facial region of the image the image features and the bloodstains do not match up. There are bloodstains in the hair of the man on the Shroud which could only have come from the sides of the face. Dr Lavoie concluded from this that when the body image was generated the Shroud had apparently deformed to a flatter draping configuration. In other words the Shroud seemed to have flattened or straightened out at the time of image formation[9].

It was largely based on this observation that the physicist and STURP researcher Dr John Jackson posed the question that must have been on the minds of other scientists, but which they were undoubtedly reluctant to articulate – is the image on the Shroud due to a process heretofore unknown to modern science[10]?

* * * * * * * * *

We have seen that the image on the Shroud bears wounds that are totally consistent with the Passion and death of Jesus Christ, as described in the Gospels. Jesus Christ not only suffered and died, according to the Gospel accounts, but three days later he rose from the dead. The question must inevitably arise – if there is a link between the image on the Shroud and the death of Jesus, might there also be a link between the formation of the image and his Resurrection?

Now, however, we move into the realm of the supernatural. Science must approach events that appear to be beyond the natural with the greatest of caution. If the image was formed by the physical phenomena surrounding the resurrection of a dead body we may never fully understand the mechanics of its formation, as we do not know, and cannot know, how exactly a body rises from the dead.

Where events occur that may be outside the known laws of science, we have to be prepared for anything. At the same time we have to preserve the highest degree of scientific detachment and look, where possible, for natural explanations and natural phenomena.

Physical explanations for the formation of the image rely on radiation of some form being the cause of the image. There are two known types of radiation that may be considered in the context of the Shroud. Electromagnetic radiation includes radio waves, infra-red light, the visible spectrum, ultra-violet light and X-rays. Nuclear radiation is the emission of various atomic and sub-atomic particles, including alpha-particles (helium nuclei, consisting of two protons and two neutrons), beta-particles (electrons), protons and neutrons. Irradiating an object frequently results in heat, which in turn can result in changes in the chemistry of the object in question. In the case of the Shroud of Turin, it has been suggested that whatever form of radiation was involved resulted in heating of the surface fibres of the cloth. This would have led to chemical changes in these fibres, causing the yellow colour observed.

The emission of radiation, of whatever form, by a dead body is obviously outside scientific experience. However in the case of the Shroud it should first be assumed that the radiation itself was of a type known to science rather than some mysterious spiritual radiation beyond human knowledge. It therefore becomes important that, in order to understand how the image was formed, we try and identify the exact form of the radiation involved, even if we cannot explain its origin.

* * * * * * * * *

Dr Jackson proposed the hypothesis that, at the time that the image on the Shroud was formed, the cloth collapsed into and through the underlying structure of the body in the Shroud. He did admit that, as a physicist, he had his own difficulties with this concept. Based on his observations of the image he further proposed that, as the body became mechanically transparent to its physical surroundings, it emitted radiation from all points within and on the surface of the body. This radiation interacted with the cloth as it fell into the mechanically transparent body, forming the body image. He also suggested that the radiation would have had to have been strongly absorbed in air. This, he suggested, could have been electromagnetic radiation in the shortwave ultra-violet region of the spectrum, which would have caused a chemical alteration of the cellulose in the cloth fibres.

He also noted several predictions arising from his hypothesis. For example, the frontal image on the Shroud should be three-dimensional as a result of the upper part of the Shroud falling dynamically through the body. The dorsal image should be one of "direct contact", as the lower part of the Shroud remained statically in place beneath the body.

If there were variations in the radiant emissions caused, for example, by initial mass density within the body, then internal body structures might appear in the image[11]. It has been noted that the Shroud image does suggest quite strongly the presence of many skeletal details, including carpal and metacarpal bones (bones in the hands and feet), some 22 teeth, eye sockets, the left femur and left and possibly right thumbs flexed under the palms of the hands[12].

In addition, the collapse of the upper part of the cloth through the body should result in the front image being visible on both sides of the Shroud, as radiation from the body would be striking both sides of the cloth.

While Dr Jackson was the first scientist to propose a mechanism as far removed from scientific experience as the cloth collapsing through a mechanically transparent body, others before him had suggested radiation emanating from the body as the cause of the image. In 1984 Prof Giles Carter of Eastern Michigan University suggested that low-energy longwave X-rays would produce the same type of superficial discolouration of cloth as is seen on the Shroud. He also noted that finger and hand bones of the man on the Shroud were visible on the photographic negative images of the Shroud. He suggested that these "images may be due at least in part to X-rays emanating from the bones in the body."[13]

Prof Carter also suggested that part of the spine of the man on the Shroud might be visible on the dorsal image on the Shroud. This has subsequently been confirmed by other investigators[14].

Another hypothesis, this time involving nuclear radiation, was developed by Dr Jean-Baptiste Rinaudo of the Centre of Nuclear Medical Research in Montpellier, France. Dr Rinaudo suggested that if the body had given off vertically polarized gamma rays (high frequency X-rays), these would have caused the release of protons and neutrons from the nuclei of deuterium (heavy hydrogen) atoms uniformly distributed over the body. According to this theory the protons would cause the image on the cloth while the neutrons would enhance the radiocarbon content throughout the cloth[15]. This theory therefore had the advantage of explaining both the image and the controversial radiocarbon dating of the cloth. However, as has been shown in the previous Chapter, there is no need to develop complex theories involving nuclear reactions to explain the carbon dating – it was quite possibly affected simply by the sample used not being representative of the Shroud as a whole.

Dr Rinaudo's hypothesis has been strongly questioned by Dr Ray Rogers, who queried the assumptions that he made and in particular questioned the probability that any significant amount of radiocarbon would be produced by the flow of neutrons[16].

The author Mark Antonacci has sought to combine features of both Dr Rinaudo's and Dr Jackson's hypotheses into what he has called a Historically Consistent Method. He suggested that the radiation involved may have been a combination of protons, neutrons, alpha particles, electrons and gamma rays, arising from some form of dematerialization of the body. This combined radiation would, he suggested, explain many of the features of both the image and of the Shroud itself that had been noted by the STURP team. It would also explain the images of coins (described in Chapter 12 of this book) and flowers (described later in this Chapter) which some observers claim to have seen on the Shroud. As with Dr Rinaudo's theory, the neutrons in the combined radiation would create new radiocarbon in the cloth. Mr Antonacci claimed that only his hypothesis could explain all the image and non-image features found on the Shroud of Turin[17].

Image Formation: Radiation Hypotheses

Dr Rogers has again been highly critical of Mr Antonacci's hypothesis. Occam's Razor is a principle of logic that Dr Rogers has used in several papers he has written. It can be stated as follows:

"The hypothesis that includes the smallest number of special assumptions has the highest probability of being closest to the truth."[18]

Dr Rogers counted sixteen such special assumptions in Mr Antonacci's hypothesis. There are certainly major points in Mr Antonacci's work that must be queried. One of these is the question of dematerialization of the body. Dr Jackson merely suggested that the body became "mechanically transparent" through some unknown process. Mr Antonacci seems to suggest that the body dematerialized with the conversion of mass into energy[19], this energy being in the form of the radiation that created the image on the Shroud. Dr Rogers has very correctly pointed out, on the basis of Albert Einstein's well-known equation on the conversion of mass into energy, $e = mc^2$, that if the body of an average man dematerialized into energy, the energy released would result in an explosion equivalent to 200 – 300 megatons of TNT – several times larger than the most powerful hydrogen bomb ever tested[20]. This would have instantly vapourised the whole of Jerusalem and a good portion of surrounding Roman Judaea.

Mr Antonacci clarified this issue to a certain extent in the end-notes of his book, *The Resurrection of the Shroud*, where he concedes that only a tiny fraction of the atoms are needed to encode the Shroud's images. He goes on to suggest that perhaps the body in the Shroud passed through a "wormhole", a bridge in space-time that permits instantaneous travel between two points in space-time. Although "wormholes" are perhaps most commonly associated in the public mind with the space adventures portrayed in episodes of *Star Trek*, there is, as Mr Antonacci has pointed out, serious theoretical physics behind the concept[21]. However the theory behind "wormholes" is related to gravitational distortions surrounding "black holes" in space. No such "wormhole" has ever been observed by physicists and Mr Antonacci fails to explain how the mouth of a "wormhole" would suddenly appear in the vicinity of the Shroud. It is very clear, and Dr Rogers has emphasized this, that Mr Antonacci has no scientific basis for invoking "wormholes" in the case of the Shroud[22].

* * * * * * * * *

At this point it is necessary to return to the supernatural and the possible link between the Shroud image and the resurrection of Jesus Christ, with all that that implies. Perhaps the Gospel accounts of the resurrection and the events that followed can offer some clues as to how the image on the Shroud *may* have been formed.

The Gospels suggest that the risen Jesus could teleport – in other words he could move apparently instantaneously from place to place regardless of the physical obstacles in the way – without the appearance of "wormholes" or other disruptions in the fabric of space-time. In John 20:19 and again in John 20:26 it is recorded that Jesus appeared suddenly among his disciples in a locked room. Luke 24:31 records Jesus as vanishing from the sight of the disciples he met on the road to Emmaus. Again, in Luke 24:36 he suddenly appears among the apostles in Jerusalem.

However, his body was substantive – it was not some form of energy or light. Luke 24:43 describes how the apostles gave him a piece of grilled fish and he ate it "before their eyes". John 20:27 records how the apostle Thomas placed his finger and hand into the wounds in Jesus's hands and side.

Clearly the body of the risen Jesus, as described in the Gospels, had physical properties beyond the knowledge of modern science. However it was a solid body and this would seem to exclude any possibility that at some point it had dematerialized in a burst of energy.

The Gospel accounts do not, however, preclude the possibility that the body of the risen Jesus became "mechanically transparent". In fact they seem to suggest it in their descriptions of how Jesus appeared and disappeared without warning. The Gospel accounts give more credence to Dr Jackson's proposed image-formation mechanism than to Mr Antonacci's.

*   *   *   *   *   *   *   *   *

One of the predictions of Dr Jackson's hypothesis that the Shroud collapsed through the body it contained was that the frontal image would be visible on both sides of the Shroud. At the time this could not be tested as the reverse side of the Shroud had been covered since 1534 with a

Image Formation: Radiation Hypotheses

backing cloth. However this cloth was removed in 2002 at the time of the restoration work carried out on the Shroud. Before a new support cloth was stitched on to the Shroud the reverse side of the Shroud was completely photographed and electronically scanned by Monsignor Giuseppe Ghiberti, a professor at Sacred Heart University in Milan and a close advisor to the Archbishop of Turin. A study of these photographs was carried out in 2003 by Prof Giulio Fanti and Roberto Maggiolo of the Department of Mechanical Engineering at the University of Padua in Italy.

Using digital image processing techniques they found that there is a very faint image on the back surface of the Shroud. The face and probably also the hands appear to be visible. However no dorsal image, such as shoulders, could be detected[23]. This finding would seem to support Dr Jackson's hypothesis to at least some degree. However continuing scientific research resulted in further hypotheses and some additional findings that directly contradicted Dr Jackson's conclusions.

At the Third International Dallas Conference on the Shroud of Turin, held in Dallas, Texas from 8 – 11 September 2005, there were several papers presented which touched directly on the subject of the formation of the image on the Shroud. Dr Mario Latendresse of the University of Montreal in Canada argued that, contrary to Dr Jackson's hypothesis, the Shroud was not flattened before the image was formed. Whereas Dr Lavoie and Dr Jackson had looked in particular at the bloodstains in the hair of the man on the Shroud as evidence of a flattening of the Shroud, Dr Latendresse suggested that bloodstains on the left arm of the image show that there could have been no longitudinal movement of the Shroud at the time of image formation. He further suggested that the bloodstains in the hair studied by Dr Lavoie, which were a key element in Dr Jackson's "collapsing cloth" hypothesis, were just that – bloodstains in the hair rather than originally bloodstains on the face of the image. There is logic in this argument if one assumes a crown of thorns – a man with a crown of thorns pressed into his scalp would, as discussed in an earlier Chapter, bleed profusely into his hair. Dr Latendresse carried out computer modeling using a linen sheet and concluded that a flattening of the upper part of the Shroud is not required to avoid major image distortions[24]. However Dr Latendresse did not attempt to address the question of how the image was formed.

This question was addressed further in a paper presented by Prof Fanti. He proposed two possible hypotheses based on corona discharge. As described by Prof Fanti:

"A corona discharge is an electrical discharge brought on by the ionization of a fluid surrounding a conductor.... a current develops between two high-voltage electrodes in a dielectric fluid, usually air, by ionizing the fluid so as to generate a plasma around one electrode."[25]

Corona discharge occurs in nature as an electrostatic discharge during thunderstorms. This is the cause of what is known as St Elmo's Fire. Also, an earthquake can cause a large electric field surrounding compressed rock layers of quartz. This can particularly occur in the presence of radon, a radioactive inert gas that makes the surrounding air a highly ionized medium.

Prof Fanti and his associates proposed two hypotheses. The first was that, as a result of an earthquake, a surface corona process in the tomb formed a complete image on the Shroud as a result of an exogenous electric field produced by the piezoelectric effect of quartziferous layers of rock surrounding the tomb. This process would have been promoted by the presence of radon, which is known to be common when there are earthquakes. Their second hypothesis was that the body itself could have been involved in the production of the corona discharge. The first hypothesis depends on purely natural phenomena while the second involves the body being involved, through some process not known to science, in the generation of the radiation field. Prof Fanti and his co-workers conducted a number of experiments with corona discharges and compared the results of these with a number of characteristics of the Shroud of Turin. He also suggested that further studies should be made in regard to a comparison of Dr Jackson's hypothesis and his second one, which involved the body itself in the generation of the corona discharge, to see if a common hypothesis could be developed which did not involve the body becoming mechanically transparent. He noted that the ultra-violet radiation assumed to exist by Dr Jackson could be caused by corona discharge. In addition the double superficiality of the body image postulated by Dr Jackson and observed by Prof Fanti himself would be compatible with his second hypothesis.

Image Formation: Radiation Hypotheses

Prof Fanti also referred to the possibility that the tomb of the man on the Shroud was in Jerusalem. He noted that quartziferous layers of rock are not present under Jerusalem and that the rock present there is unlikely to generate piezoelectric effects. This would seem to rule out his first hypothesis, involving purely natural phenomena, if it is assumed that the events described took place in Jerusalem, unless there was a particularly high concentration of radon present. He noted that more detailed study is required in this area.

In the conclusion of his paper Prof Fanti conceded that his purpose was not to completely explain how the body image was formed on the Shroud, but rather to propose an energy source based on corona discharge and its collateral effects, with a view to possible further studies in this area[26].

A third paper of interest at the Dallas Conference was presented by Dr Lavoie. This was a study of both the frontal and dorsal images on the Shroud from a forensic medical perspective. In the second part of his paper he looked in particular at the back image on the Shroud and suggested that there is no flattening of the image, as would be expected from a body lying on a hard surface. On the contrary, in Dr Lavoie's view the image on the Shroud resembles more what would be expected from a man in an upright position[27].

In his book, *Resurrected*, Dr Lavoie recalled meeting the Shroud historian Dorothy Crispino in Rome at the time of a Shroud Symposium there in 1978. While looking at a life-size picture of the back image in a darkened room, Ms Crispino remarked to him, "You know, one thing that this light does is bring out the roundness of the muscles, as if the person were right there in front of you."[28] The muscles of the backs of the legs in the image do indeed appear rounded rather than flattened.

Dr Lavoie went on to argue, and demonstrate, in his book that the image of a body in the horizontal position would have its hair falling back to the ground rather than straight down to its shoulders, as is the case with the image on the Shroud[29]. However, it must be borne in mind that if the man in the Shroud was indeed a crucified first-century Judaean, his hair is likely to have been matted with dirt and blood. It would not have fallen back in the same way as would the clean hair of a modern person lying supine.

Dr Lavoie also noted that, although upright, the man could not have been standing, as the soles of his feet are not correctly placed for this. The man, rather, seems to Dr Lavoie to be suspended in mid-air in a vertical position[30].

A possible explanation for Dr Lavoie's observations is that the body was still in a state of rigor mortis when the image was formed. As we have seen rigor mortis could well have started before the body was placed in the Shroud – rigor mortis would have set in very quickly in the event of a violent death[31]. Rigor mortis becomes complete in about 18 to 36 hours, it remains for about 12 hours and then leaves the body over a period of about a further 12 hours[32]. Thus rigor can affect the body for up to 60 hours, although this is variable. If some degree of rigor mortis had set in to the body before it was placed in the Shroud and rigor continued until the time of image formation, it is quite conceivable that there would have been no flattening of the body as a result of lying on a hard surface. The position of the feet on the image is also typical of rigor mortis. There are therefore possible explanations for Dr Lavoie's observations regarding the image that do not require the body to be in a vertical position. His suggestion does, however, add an interesting new possible dimension to the problem of how the image was formed.

* * * * * * * * *

Quite clearly, if the image on the Shroud is the product of some form of radiation, the image formation mechanism was extremely complex. If the body in the Shroud was itself somehow involved in the emission of a radiation field, this goes beyond the bounds of current scientific knowledge or experience and could possibly be termed supernatural. The exact processes involved could therefore defy our understanding.

The idea that the body may have been somehow suspended in mid-air at the time of image formation adds yet another dimension of complexity. It is the view of Dr Lavoie, who suggested the idea, that this was an integral part of the miraculous resurrection of Jesus Christ. He also commented that:

"Science can only do so much, and so far, it tells us that the image is a wonder that remains unexplained."[33]

One image formation hypothesis that requires an upright body is the hypothesis that the Shroud is a forgery involving some form of early photographic process. In 1988 Professor Nicholas Allen of the Port Elizabeth Technikon in South Africa carried out experiments to see if he could replicate the Shroud image photographically using substances available to a person of the Middle Ages. He constructed a *camera obscura* and made use of rock crystal, silver salts and salts of ammonia to create an image of a plaster cast of a human figure on a shroud-like cloth. Prof Allen succeeded in creating an image that bears a number of the characteristics of the image on the Shroud. For example, there was a straw-yellow discolouration of the upper fibrils of the linen material; the image gave the appearance of being a photographic negative; there was no directionality in the image; it had thermal, water and a certain degree of chemical stability. Prof Allen then photographed his image using black-and-white film and viewed this in negative. The result was a naturalistic positive photograph of a naked man laid out as in death, which, like the image on the Shroud itself, could not be interpreted as the work of an artist[34].

Prof Allen's work was imaginative, but while it showed that a mediaeval scientist *may* have been able to create an image on a cloth using photographic means, it suffered from major flaws as an explanation of the image on the Shroud. Firstly, the image on the Shroud has been clearly shown by forensic scientists to be that of a dead man who died a particularly gruesome death. The possibility of a mediaeval photographer using a dead body covered with the marks of crucifixion seems extremely unlikely – the photographic exposure time of several days in bright sunlight would have led in itself to considerable decomposition of the body. Prof Allen himself used a plaster cast of a body. It has also been clearly shown that the image on the Shroud was formed *after* the bloodstains had been transferred to the Shroud. A mediaeval forger would have not only had to sprinkle his cloth with human blood before starting the photographic process, but would have had to place the blood in exactly the right places to be fully consistent with the subsequent human image. Prof Allen's work was carried out in the context of the Shroud being of mediaeval origin – Dr Rogers has since shown that the Shroud is considerably older. A mediaeval photographer would have had to use a cloth that was already old in his time. There are many other medical and scientific reasons why the hypothesis of the Shroud being the product of a 14th or 15th century photographer does not stand up to detailed scrutiny. One point particularly worth bearing in mind is that no mediaeval forger could or would have produced a *negative* image on a cloth that could only be seen as a 'positive' using technology that would not be available until five hundred years in the future.

A more fanciful theory, based also on the use of mediaeval photography, was that of Lynn Picknett and Clive Prince in their book, *Turin Shroud: in whose Image*. They suggested that the image had been produced by photographic means by Leonardo da Vinci in the late fifteenth century. The same scientific objections to an image-formation hypothesis based on Prof Allen's work apply equally to this hypothesis. In addition, as has been shown in an earlier Chapter, the historical claims of these authors with regard to the whereabouts of both the Shroud and da Vinci at that time do not match historical fact[35].

Mediaeval photography is not a valid explanation for the formation of the image on the Shroud. This leaves the subject still surrounded by mystery. No image-formation hypothesis based on radiation effects has as yet provided a satisfactory explanation for the image. Further studies are undoubtedly required, but there may possibly be alternative explanations.

* * * * * * * * *

Moving right into the realm of the mystical and the supernatural, there is another possible explanation for the formation of the image on the Shroud – that it was formed by a mysterious radiation not known to science.

Eastern Orthodox Christianity tends to place more emphasis on the mystical than does Western Christianity. For some Orthodox mystics, the culmination of mystical experience is the vision of Divine and Uncreated Light, described by the Byzantine mystic St Simeon the New Theologian at the turn of the tenth century as "…fire truly divine, fire uncreated and invisible, without beginning and immaterial." It is this light, the Orthodox believe, that surrounded Jesus at his Transfiguration on Mount Thabor and which is the Light of the Godhead Itself[36].

Image Formation: Radiation Hypotheses

An Orthodox mystic would find nothing extraordinary in the thought that the body of Christ, the Son of God, may have emitted the Divine and Uncreated Light at the moment of Resurrection, and that this Light, physically beyond our understanding or measurment, created the image on the Shroud.

Chapter 18 – Image Formation: Contact and Chemical Hypotheses

The Shroud researchers of the early twentieth century, such as Paul Vignon and Jean Volkringer, looked at ways in which the image might have been formed either by contact between the body and the Shroud or through chemical reactions involving the body and the Shroud. Radiation is very much an area of study for physicists and the early Shroud researchers tended to be in areas of study related to the medical profession. Paul Vignon was a biologist; Jean Volkringer was a pharmacist. The first physicist to enter the study of the Shroud and to develop radiation-based hypotheses of image formation was Dr John Jackson in the early 1970s. The early researchers also assumed that the image would have been formed by entirely natural processes. It was Dr Jackson who first suggested that image-formation may have been the result of a process unknown to modern science.

Today different hypotheses of image-formation tend to reflect the different disciplines, and beliefs, of the scientists proposing them. Radiation-based hypotheses are the speciality of physicists; chemists, such as Dr Ray Rogers, look for possible ways in which the image could have been formed through chemical reactions. Others, sceptical laymen as well as scientists, have sought to explain the image using the supposition that it was deliberately man-made, as a hoax or forgery. Nobody today seriously suggests that the image is a painting – the work of STURP effectively ruled out that hypothesis, despite the claims of Dr McCrone – but other possible techniques, some more ingenious than others, have been proposed from time to time.

One such sceptic was a one-time stage magician and amateur detective, Joe Nickell. He experimented with a possible technique of image formation

Image Formation: Contact and Chemical Hypotheses

which involved conforming wet linen to a bas-relief, while impressing all the relief's features on to the cloth. After the cloth dried he used some cotton covered with cloth to rub powdered pigment on to the impressions left on the linen. He went on to suggest that the powdered pigments used on the Shroud itself were iron oxide, myrrh or aloes. He claimed that his powder-rubbing method produced a superficial image. However experiments by STURP scientists using this method resulted in powder particles falling through the weave of the cloth and accumulating on the reverse side. There are a number of other reasons why Mr Nickell's technique is not valid. VP-8 photography of the image created by powder-rubbing does not show a true three-dimensional image. A uniform application of powder, resulting in a uniform image, has been shown to be unachievable. There is no evidence of such a powder or any possible by-products of a powder in the material of the Shroud, which has been subjected to detailed microscopic examination. There is no sign in the Shroud material of tensions that would inevitably result from a wet cloth being moulded to a bas-relief. There is no historical evidence of powder-rubbing being used by artists before the nineteenth century. Mr Nickell has not produced any reasonable explanation for the blood-marks on the Shroud, and indeed he has never submitted his own experimental cloth for serious scientific examination[1].

Prof Randall Breese and Dr Emily Craig of the University of Tennessee proposed that a method similar to Mr Nickell's had been used to produce the image on the Shroud. Their suggestion was that an artist could start by drawing, freehand, three-dimensional information of a face and body on to a paper or similar surface. A cloth would then be laid over the drawing and the image transferred on to the cloth from the paper rubbing the back of the cloth. The cloth would then be heated to bring out the image. This method did result in a visible image on the cloth but it did not show the diffuseness of the image on the Shroud. In addition such a method would inevitably result in a transfer to the cloth of some of the pigment or powder used to draw the image on the original surface. No such pigment or powder has been detected on the Shroud. Finally, this method would not account for the blood marks and wounds that appear on the Shroud[2].

There have been a number of similar attempts to try to use contact methods to create an image similar to that on the Shroud. The small town of Moscow, Idaho in the United States gained fleeting news coverage in early 2005 when Nathan Wilson, a Fellow in Literature at a small Christian Reformed college there, claimed to have created a remarkable likeness of the Shroud image. Mr Wilson had an artist friend paint a picture of Jesus in oils on a pane of glass. He then placed a linen cloth on the glass and left the cloth and the glass to bleach in the sun on his college roof. It was subsequently pointed out that distortion-free glass did not exist until the seventeenth century and that the Shroud of Turin presents both frontal and dorsal images with no lines indicating that it had been laid on a pane or panes of glass[3]. It seems unlikely that any serious scientific analysis or study of Mr Wilson's efforts has been made.

Other suggestions have included pressing a stretched cloth over a heated bas-relief or a full-size three-dimensional hot statue. The heated bas-relief method was first proposed in 1961. It had particular interest because the image on the Shroud does seem to have many of the physical properties of a light scorch. Three STURP scientists, including Dr Jackson, tested this hypothesis by heating a full-size bas-relief model of a face and stretching it over a linen cloth. The resulting image lacked the high resolution and focus of the Shroud image and VP-8 relief images showed clear distortions. Another failing of this hypothesis is that there is clear evidence that the blood images were transferred to the Shroud before the body image. If the body image had been transferred through contact with a hot surface, the bloodstains on the Shroud, being already on the cloth, would have to show evidence of thermal discolouration or fusing. There is, however, no such evidence on the bloodstains on the Shroud.

The hot statue hypothesis has the failing that the heat from such a statue would radiate equally in all directions. The image on the Shroud was formed by transference of the image information along vertical, straight-line paths. Heat from a statue would produce a blurred image rather than the focused image of the Shroud[4].

No attempt to produce a Shroud-like image using any physical contact method has produced an image that comes close to the one on the Shroud. Every attempt or hypothesis has failed to meet crucial scientific criteria for the Shroud image. Although sceptics continue to claim that the Shroud is a mediaeval forgery, none has been able to produce a workable hypothesis that stands up to scientific scrutiny as to how the image was created by the mediaeval forger.

Image Formation: Contact and Chemical Hypotheses

* * * * * * * * *

Not only sceptics have looked for ways in which the image on the Shroud may have been created using natural methods that can be easily explained and replicated. Biochemists and chemists have sought to explain the image in terms of natural processes involving their own scientific disciplines.

The American biochemist Dr Leoncio Garza-Valdes, who suggested that the result of the carbon-dating of the Shroud had been distorted by a bioplastic coating on the material, has sought to explain the image in biochemical terms. It is his belief that the image was formed by a relative deposit of bioplastic coating. The areas that look darker are the areas that had more contact with the body in the Shroud; as a result the bacteria in those areas grew faster. The concentration of sweat, salt, oils and blood in those areas promoted the faster growth of the bacteria.

Bacteria, as Dr Garza-Valdes pointed out, grow very slowly and under particular conditions. He felt that if the Shroud had been stored away for a period of several centuries, as suggested by Ian Wilson in his book *The Turin Shroud*, in suitable conditions of temperature and humidity, the image would have formed gradually over this period as a result of bacterial activity[5].

Dr Garza-Valdes' hypothesis of a bioplastic coating has not been supported by any other scientific studies and has been effectively discredited by the observations of Dr Ray Rogers and Anna Arnoldi referred to earlier, in Chapter 16. Dr Rogers and Ms Arnoldi specifically mentioned that "several different analytical methods have failed to detect (on the Shroud material) any of the elements other than carbon, hydrogen and oxygen that are necessary for the growth of organisms."[6] It must therefore be concluded that his theory of formation of the image by bacterial means is equally invalid.

Dr Rogers has been a prolific researcher and writer on the subject of the Shroud and how the image was formed. A member of the 1978 STURP team, he was the co-author in 1982 of a paper summarising the STURP results[7]. In the years immediately prior to his death in 2005 he published several papers giving a view of the problem of image formation from a chemical perspective. Before looking in detail at possible chemical explanations for the image he reached the conclusion that radiation could not have caused the formation of the image. He noted from the STURP studies that the image colour is only on the surface of the image fibres and went on to state that the observations of the STURP team prove that "only radiation-induced reactions that colour the surfaces of fibres without colouring the cellulose can be considered."[8]

His next observation was that the medullas – tubular voids in the centres of linen fibres – in the Shroud cloth do not show any colouration or charring, except where the fibres were scorched during the fire that damaged the Shroud in 1532. He then made a critical observation:

"If any form of radiation (thermal, electromagnetic or particle) degraded the cellulose of the linen to produce the image colour, it would have had to penetrate the entire diameter of a fibre in order to colour its back surface. Some lower fibres are coloured, requiring more penetration. Radiation that penetrated the entire 10 – 15 micromillimetres diameter of a fibre would certainly colour the walls of the medulla. All image fibres show colour on their surfaces but not in the medullas…. No radiation hypothesis alone can explain how the entire outer surface of image fibres could become coloured without colouring the inside and the medulla."[9]

Dr Rogers also gave consideration to the corona discharge theory of image formation – described in the previous Chapter. He attempted to produce colouration of fibres using corona discharge but was unable to do so[10].

Dr Rogers stated clearly that a logical hypothesis for image formation must accept the laws of physics and chemistry and explain all of the STURP observations, of which he listed thirteen. From this he concluded that no single, simple hypothesis will be adequate to explain all of the observations made on the Shroud[11].

Dr Rogers went on to suggest that the image may have been the result of a complex chemical process called a Maillard reaction. These are fast, colour-producing chemical reactions that involve several reaction paths and products. Starting with a complex mixture of carbohydrates such as those in linen there could be thousands of products. In Maillard reactions, amino groups (which contain nitrogen) are needed to react with the carbohydrates – chemicals containing amine reactants would be produced very quickly by the decomposition process of a dead body. However, some of the most important products of such reactions do not contain nitrogen, and this would be consistent with the STURP observation that there were no nitrogen compounds in image areas of the Shroud.

He also suggested that a chemical called "soapweed", which was used to wash ancient cloths before they were used, could have been an integral part of an image-forming Maillard reaction.

He then made a formal statement of a chemical hypothesis for image formation:

"The cloth was produced by technology in use before the advent of large-scale bleaching. Each hank of yarn used in weaving was bleached individually. The warp yarns were protected and lubricated during weaving with an unpurified starch paste. The finished cloth was washed in *Saponaria officinalis* (soapweed) and laid out to dry. Starch fractions, linen impurities and *Saponaria* residues concentrated at the evaporating surface. The cloth was used to wrap a dead body. Ammonia and other volatile early amine decomposition products reacted rapidly with reducing saccharides on the cloth in Maillard reactions. The cloth was removed from the body before liquid decomposition products appeared. The colour developed slowly as Maillard compounds decomposed into final coloured compounds."[12]

In carrying out initial tests of his hypothesis using Maillard reactions in the laboratory, Dr Rogers observed that fibres were coloured a golden-yellow similar to the colours on the Shroud; that the coating of the Maillard products was all on the outside of the fibres; and that there was no colouration in the medullas[13].

Dr Rogers returned to the question of radiation effects on the Shroud of Turin in a paper published in 2005. Again he emphasized that image formation proceeded at normal (ie atmospheric) temperatures and that formation of the image colour did not require neutrons, protons, high-energy photons or mesons. He then particularly addressed the hypothesis that the formation of the image had involved a corona discharge. He noted that any corona in air will result in oxidization of the material of a cloth in the vicinity. The surface of flax fibres will be eroded. Controlled experiments were carried out by a manufacturer of equipment for plasma treating textile, APJet of Santa Fe in New Mexico. In these experiments the fibres of a cloth produced following methods used in the Near East in Roman times were oxidized and eroded without any charring. The plasma, which is electrically neutral, completely penetrated the pores of the cloth. These effects on the linen sample used are not observed in image fibres from the Shroud of Turin. It was therefore Dr Rogers' conclusion that corona discharges and/or plasmas made no contribution to image formation. Indeed, he expressed his belief that all radiation-based hypotheses for image formation will ultimately be rejected[14].

* * * * * * * * *

Yet there is another peculiar phenomenon on the Shroud that is difficult to explain by a chemical process. In 1983 a German physics teacher, Oswald Scheuermann, claimed to have observed flowerlike patterns around the face of the man on the Shroud[15]. Two years later Dr Alan Whanger, Professor Emeritus at Duke University, North Carolina, reported that he had also observed the images of flowers on the Shroud, while studying photographs of the Shroud. Dr Whanger and his wife, Mary, continued with studies of enhanced photographs taken by Enrie in 1931 and eventually claimed to have identified 28 species of plant from images on the Shroud. All of these plants grow in Israel. The floral images found by the Whangers included hundreds of flowers, many of them around the head of the man on the Shroud and others extending down to the level of the waist, including some described as bouquets[16].

In order to confirm their observations, Dr and Mrs Whanger approached an Israeli botanist, Dr Avinoam Danin, Professor of Botany at the Hebrew University of Jerusalem. Dr Danin indeed confirmed the Whangers' observations and noted that "while the images are of slightly wilted flowers rather tightly clustered together, many of them are quite identifiable even though they are faint, partial and of low contrast". He also noted that Oswald Scheuermann had produced similar images from flowers in experimental studies with corona discharge[17].

Dr Danin also noted that some of the species identified correlated with the pollen found on the Shroud and identified by Dr Max Frei. In particular Dr Danin identified *Gundelia tournefortii*, a thorn, near the anatomical right side of the head of the man on the Shroud. Dr Frei had found 45 pollen grains of this plant in the Shroud. *G. tournefortii* grows only in the Near East and blooms in Israel from February to May. Another plant identified by Dr Danin was *Zygophyllum dumosum Boiss*, which is found only in Israel, Jordan and the Sinai desert. He claimed to have identified the image of this plant on the Shroud itself, using binoculars during the exposition of the Shroud in Turin in 1998, as well as to have observed the images on different sets of photographs taken by different photographers, thus indicating that the images are not peculiarities of a particular set of photographs[18].

Dr Danin concluded that:

"Using my data base of more than 90 000 sites of plant distribution, the place that best fits the assemblage of the plant species whose images and often pollen grains have been identified on the Shroud is 10 – 20 km east and west of Jerusalem. The common blooming time of most of these species is spring – March and April."[19]

Dr Whanger has also noted that flower images similar to those on the Shroud have been portrayed in artistic images of Christ similar to the image on the Shroud. He referred specifically to a fourth century portrait of Christ in a Roman catacomb, the Christ Pantocrator icon of St Catherine's Monastery at Mount Sinai, and images of flowers on representations of the Shroud on Byzantine gold coins of the Emperor Constantine VII in 950, after the Cloth of Edessa had been brought to Constantinople[20].

Such images cannot be explained by a Maillard reaction, as this relates in the context of the Shroud to the decomposition products of a human body, which are obviously not found in flowers. Dr Rogers does not seem to have commented specifically on the work of Dr Whanger and Dr Danin, but in one of his papers he did make the following remarks:

"With regard to other images on the Shroud (the "coins" on the eyes as well as the flowers), few of us can see them. "I think I can see" is not a substitute for an observation, and observations must be confirmed....

Your mind tries to make sense out of any "patterns" your eye can see. Psychologists have a lot of effort invested in studying such phenomena. It is dangerous to build a scientific theory on such shaky foundations. Your mind tends to see what it expects and/or wants to see."[21]

There is clearly no consensus among scientists as to how the image on the Shroud may have been formed. No single hypothesis or explanation seems to account for it. One set of observations seems to rule out all theories based on radiation processes; other observations seem to support such theories. The conclusions of STURP regarding the complexity and inexplicability of the process of image-formation are as valid today as they were in 1978. However, to suggest that the image must therefore be nothing other than miraculous and hence beyond possible explanation by science also begs the question. Whatever the cause of the image, the process by which it was formed should be explicable in scientific terms. The fact that no scientific explanation has yet been found does not automatically mean that there is not one.

Chapter 19 – Preservation, Restoration and the Future for Research

The history of the Shroud has been one of controversy. During its known history it has been the focus of personal ambition and greed, faith and scepticism, and, more recently, scientific argument. Nothing changes.

The restoration of the Shroud carried out in 2002 and announced at a press conference in Turin on 19 September 2002 has proved as controversial as any previous event in the Shroud's history.

The Shroud of Turin Research Project (STURP) was formally disbanded in 1993[1], leaving a void in the area of formal scientific research on the Shroud. The carbon-dating of 1988 and its aftermath had overtaken and over-shadowed the work done by STURP. Proposals for further scientific testing of the Shroud itself had lost momentum. At this point a change of emphasis occurred, away from scientific testing of the Shroud to how it could best be conserved. In 1992 a conservation committee was set up by the then Archbishop of Turin, Cardinal Giovanni Saldarini. This consisted of five textile experts from the USA, England, Switzerland and Italy. The objective of this committee was to:

"…set in motion a complete cycle of studies on the fabric and on the best conditions for its conservation…."[2]

The committee produced a report in February 1996 which made a number of suggestions and recommendations for the best conservation of the Shroud. One of these was that it was necessary to study in depth the question of the removal and substitution of the backing Holland cloth which had been put in place in 1534, as well as the possible removal and replacement of the patches stitched on to the Shroud at the same time. It was also agreed that, in future, the Shroud would be laid out flat and in a horizontal position, free of stress and tension, rather than being kept rolled up around a wooden cylinder. This report was submitted to the Papal Custodian of the Holy Shroud, the Archbishop of Turin, and also to the Vatican[3].

The composition of the committee changed during the second half of the decade and by 1999 only one of the original members remained on it, Dr Mechthild Flury-Lemburg, the Swiss textile expert known for her work in the restoration of ancient fabrics. Other scientists, in other disciplines, had joined the committee, now known as the "Commission for Conservation", including the former STURP scientist Prof Alan Adler. (Prof Adler died the following year.)[4]

The Shroud was publicly displayed in 1998, to mark the centenary of Secondo Pia's first photographs of the image, and again in 2000 to mark the Holy Year. After this second exposition the Shroud was placed in a new airtight case in an atmosphere consisting primarily of the inert gas argon. Before being sealed in its new case, the Shroud was again officially photographed using both traditional and digital photography. In addition, the reverse side of the Shroud was partially examined using a scanner after the Shroud had been temporarily unstitched from its backing cloth along its perimeter. It had previously been photographed using fibre optics during the 1978 STURP examination. This examination confirmed the existence of bloodstains on the reverse side of the cloth, although this had already been established by the STURP photographs[5].

This examination of the reverse side of the Shroud caused some upset in the international community of Shroud researchers. There had been no opportunity for peer review or any consultation at all among scientists outside the Commission for Conservation. The STURP photographer Barrie Schwortz was particularly critical. He emphasized his feeling that it was wonderful that 21st century technology was being applied to Shroud research, but strongly criticized the Turin authorities for taking an opportunity to gather new data without considering many research proposals submitted by leading researchers around the world, as well as for applying modern technology to questions that had been answered years previously[6].

As Mr Schwortz pointed out, the tests carried out and conclusions reached by the 2000 examination merely repeated work done by the STURP team in 1978.

At around the same time as the examination of the Shroud's reverse side was being conducted the Commission for Conservation apparently reached a decision that a major restoration of the Shroud should be carried out by removing the patches and the backing Holland cloth, as had been suggested in 1996. The recommendation of the Commission was endorsed by Cardinal Severino Poletto, the new Archbishop of Turin, and subsequently approved by the Secretary of State for the Vatican, Cardinal Angelo Sodano[7].

The restoration itself was carried out between 21 June and 23 July 2002. It was overseen by Monsignor Giuseppe Ghiberti of the Archdiocese of Turin, who was also a member of the Commission for Conservation. The technical work was directed by Mme Flury-Lemburg. The first phase of the restoration involved removing the patches from the 1532 fire. Once the patches were removed, the burn material from the fire was micro-vacuumed and preserved in some 200 small glass vials. Next the Shroud was turned over and the Holland backing cloth removed. This was the first time in more than 450 years that the back of the Shroud had been exposed.

This was followed by scientific studies using both scanning equipment and photography. Both sides of the Shroud were digitally scanned and the images stored on a computer. However, at some point procedures were introduced that were to arouse the ire of the predominantly American scientific community that had been involved in studies of the Shroud for some 30 years. It was decided that not only should all the burn fabric be snipped away, but that the fire-darkened threads extending out into the weave from the burn holes should also be removed, either with tweezers or a scraping method. This was done not only to the burn holes from the 1532 fire, but also to the much older "poker holes"[8].

It has been suggested that lead weights were then hung from the Shroud as part of an effort to remove wrinkles. This was, however, vigorously denied by Mme Flury-Lemburg at the Third International Dallas

Conference on the Shroud of Turin in September 2005. She also stated that the Shroud had not been steam-cleaned or ironed to remove wrinkles – other accusations that had been made[9].

Following the scientific studies the Shroud was stitched down on to a new backing cloth. This was a piece of linen, some 50 years old, that had belonged to Mme Flury-Lemburg's father[10].

News of the restoration appeared in the Italian press in August 2002 and the Shroud in its new condition was displayed to a select, invited group of Shroud researchers from both Europe and America on 20 September 2002. A press conference to announce the restoration was held in Turin the following day.

There followed a welter of criticism about the manner in which the restoration had been planned and carried out and about the scientific techniques and methods used. The criticisms can be summed up in the words of Barrie Schwortz:

"Most researchers objected to the lack of any scientific involvement in what ultimately became more of a "makeover" than anything else. And considerable scientific data was disturbed or destroyed in the process, not to mention the fact that the Shroud was put at risk under 32 days of UV light exposure and the polluted air of Turin."[11]

* * * * * * * *

In order to make an accurate assessment of the restoration and its effects, the points raised in Mr Schwortz's summary need to be examined in detail.

In 1999 the names of the members of the Commission for Conservation were published. They were:

- Monsignor Giuseppe Ghiberti, Vice-President of the Diocesan Commission (in Turin) for the Exhibition of the Shroud.
- Prof Piero Savarino, an organic chemist and scientific advisor to the Archbishop of Turin.

- Prof Alan Adler, an inorganic chemist and former STURP member.
- Engineer Gian Luigi Ardoino, a civil engineer in Turin.
- Prof Pierluigi Baima-Bollone, a forensics specialist and President of the Centro Internazionale di Sindonologia in Turin, who had worked with the STURP team in 1978.
- Prof Bruno Barberis, a mathematician and Director of the Centro Internazionale di Sindonologia.
- Prof Karlheinz Dietz, a historian at Warburg University in Germany.
- Mme Methchild Flury-Lemburg.
- Prof Silvano Scannerini, a microbiologist at Turin University.
- Prof Paolo Soardo, a member of the National Electrotechnical Institute of Turin.
- Prof Carla Enrica Spantigati, Superintendent of Cultural Heritage in Piedmont.

Prof Adler, the only member of the original STURP team on the Commission, died suddenly in June 2000[12].

What is immediately clear from the composition of the Commission is that scientists were in a minority. It was not a scientific commission or body. It did invite proposals for scientific research on the Shroud in August 2000. Apparently a number of proposals were received by the Commission but these were never acted upon and none of those who submitted proposals were consulted about the tests and examination of the Shroud carried out by the Commission in November that year[13].

There was no peer review of the results of those tests and similarly there was no wide scientific consultation prior to the restoration of the Shroud in 2002.

It is easy to blame the Commission for being unscientific or parochial. This is probably unfair to a group of people who were undoubtedly distinguished in their own rights and in their own fields. The difference in approach between the Commission and the scientific community which has been highly critical of the Commission can best be explained by trying to understand how these two conflicting bodies perceive the Shroud.

To the Catholic Church, including the diocesan authorities in Turin who obviously had a major influence on the Commission, the Shroud is primarily a religious relic, an object of religious veneration. Relics have been described by the Catholic Church as "signs which can be useful to spread devotion to holy men and women all over the world"[14]. The scientific nature or even authenticity of a relic is not of prime concern. What is of concern is that the relic is a continuing focus of devotion for its adherents. Even when a relic is clearly of doubtful authenticity, if it continues to be venerated and to have religious value it remains important.

Seen in this light, preservation and conservation of the Shroud are of far greater importance than scientific study. Even the controversy over the radiocarbon dating of the Shroud would be of lesser importance to the Church authorities than its continuing value as a major relic of Christ's passion and death. This is borne out by the contents of a letter written by Fr Peter Rinaldi, the Catholic priest and Shroud researcher who had been involved with STURP from its inception, immediately after the results of the radiocarbon dating were announced in 1988. He wrote:

"The Church, which for centuries has venerated the Turin Shroud much as it did other sacred icons, never officially stated that the Shroud is actually the burial cloth of Jesus. It could hardly have done so without the support of historical and scientific proofs. It was exactly in order to determine the true origin of the Shroud that Church authorities agreed it should be examined by a group of scientists. Now that the verdict of the carbon-14 test is in, the Church will not question the results. Any disagreement among the experts on their validity will have to be settled by them.

"Whatever the scientists' decision, the Shroud will continue to be a highly revered object. It is unquestionably the most impressive visual representation of Christ's sufferings, the only one of its kind in fact known to exist....

"....the fact is, aside from what the experts have said or may yet say about the Shroud, it will continue to be, in the words of Pope John Paul II, 'a unique and mysterious object, its image a silent witness to the passion, death and resurrection of Christ.' "[15]

To the scientific community on the other hand, the Shroud is an object of major scientific and historical interest. If it can be shown to be a genuine shroud of 1st century Jewish origin, then it has much to tell scientists and historians about a broad range of subjects – the influence of the image on iconography, the realities of a Roman crucifixion, the origins of Christianity itself. It is inevitable that scientists would seek to understand the nature of the image and how it was formed, as this could give them insights into unexplored areas of physics and chemistry.

It is also of archaeological interest as, if genuine, it records a historical event that is at the centre of a major world religion.

For these reasons it is of interest to scientists from a range of disciplines – medicine, physics, chemistry, archaeology and all of the others that have been involved in the study of the Shroud since the start of the twentieth century. As I have previously emphasized, from the standpoint of science it cannot be the preserve of one scientific discipline. Similarly, a scientific viewpoint would and does argue that the Shroud cannot be the preserve of one religious group.

An appreciation of these alternative perceptions of the Shroud does lead to an understanding of what Barrie Schwortz described as "the lack of any scientific involvement in what ultimately became more of a 'makeover' than anything else."

* * * * * * * * *

Of greater concern must be the suggestion that the restoration of the Shroud has led to the irretrievable loss of valuable scientific data. The sternest critic of the restoration in this regard has been Prof William Meacham, the University of Hong Kong archaeologist who expressed reservations about the results of any carbon-dating of the Shroud being affected by contaminants in the material[16].

Mme Flury-Lemburg, the technical director of the restoration team, was of the opinion that "all efforts regarding the conservation of the Shroud must… aim at the preservation of the image". This of course is unarguable. She was also of the view that the image on the Shroud was the "product of some kind of oxidation". As has been detailed in the previous two chapters of this book, there are a number of scientific hypotheses regarding the formation of the image, none of them proved and all of them complex. This assumption by Mme Flury-Lemburg is definitely an over-generalisation by somebody who is not an expert in the field of either physics or chemistry. The Conservation Commission based its decision to proceed with the restoration on the basis that it was "urgent to find effective measures to counteract the oxidation process"[17].

Specifically, the Conservation Commission felt that there could be a danger of accelerated oxidation of the Shroud resulting from residues of burnt material in the areas of the holes caused by the 1532 fire. Such residues would be trapped between the Holland cloth and the patches. Such accelerated oxidation of the Shroud would result in the disappearance of the image. This conclusion was a key factor in the decision to remove the Holland cloth and the patches and to remove what residues could be found.

Prof Meacham took issue with this conclusion. It was his view that the residues of burnt material are merely inert carbon, that there is no evidence that the carbonized edges of the burnholes had accelerated a process of oxidation and that there is no known chemical process by which they could[18]. In other words the whole scientific basis on which the work was carried out was, in his opinion, invalid.

The manner in which the work was done has also been subject to strong criticism. Prof Meacham has described the procedures used during the restoration as being radically invasive and diametrically opposed to modern conservation practice[19].

Another archaeologist, Paul Maloney, has reiterated a point first made by Prof Meacham some years ago, that portions of the study of the Shroud were like an archaeological excavation. One of the features of an archaeological site is often its stratigraphy[20]. Samples from a site are taken from different strata levels and the position of a particular sample in the strata will tell the archaeologist much about the history of both the sample and the site. Both Prof Meacham and Mr Maloney, as archaeologists, are critical of the fact that the burn material vacuumed from the Shroud appears to have been stored without any record of its stratigraphy, or position in the material of the Shroud. Prof Meacham in particular has emphasized that the removal of charred material should have been done in a systematic manner, with proper controls and comprehensive recording. Without this the material removed is of little use for further scientific study[21].

It must be said, however, in fairness to the restorers that all the material removed from the Shroud has been stored with a view to future scientific examination, even if the method of its removal and storage has met with scientific criticism.

Concern has also been expressed that the restorers worked on the Shroud without wearing gloves, thereby possibly contaminating the material with skin oils and fresh DNA[22], and that, as previously mentioned, the Shroud was put at risk through unnecessary exposure to ultra-violet light in particular[23].

Dr Ray Rogers has also expressed his concern about the possible loss of chemical information as a result of the restoration. He observed that "the persons involved in the restoration of June and July 2002 did not appear to be familiar with previous scientific observations" – a particularly relevant remark in view of the untimely death of Prof Adler and the resulting lack of any significant continuity between the work of STURP and the work of the Conservation Commission. He also noted that those involved in the restoration did not consult chemists with different areas of experience or chemically-oriented textile conservers. He regretted that the restoration had destroyed significant chemical information[24].

The major concern of the scientists is that not only has an opportunity been lost for further scientific study of the Shroud, but that a lot of material that could have been used in future scientific studies has either been damaged or irretrievably lost. There is also concern that irreparable damage may have been done to the Shroud itself.

* * * * * * * * *

"What's done is done." The words of Lady Macbeth are as applicable to the restoration of the Shroud as they were to her husband's bloody deeds. Or, as Barrie Schwortz put it on his website, www.shroud.com, following the restoration:

"...we have to remember that the work has been completed and is irreversible. And no amount of debate or recrimination can change that. For better or for worse, the deed is done and we will all have to live with the consequences."[25]

The emphasis must now be on future studies on the Shroud. Dwelling on past difficulties will achieve nothing. Future research will depend on co-operation and understanding between the scientific community, largely based in the United States of America, and the religious authorities in Rome and Turin. A start to such co-operation was made at the Third International Dallas Conference on the Shroud of Turin, held in Dallas, Texas in September 2005 and hosted jointly by AM*STAR, the American Shroud of Turin Association for Research, and the Centro Internazionale Di Sindonologia from Turin. Papers were presented by both American and European researchers and there was frank discussion on the subject of the restoration. However, signs of some distrust remained. Questions from the floor were not allowed, apparently to avoid any embarrassment to the delegates from Turin, and at the concluding banquet there was a clear separation between the AM*STAR and Centro delegates. To quote one commentator, "No détente in Dallas."[26]

As well as different perspectives by the religious and scientific communities, there also appears to be a transatlantic divide to be overcome. While the restoration was totally the work of Europeans – there had only been one non-European on the Conservation Commission, Prof Adler, who died before the restoration was undertaken – the criticisms of the restoration came primarily from researchers based in the United States or elsewhere outside Europe. The heavy-handed approach of some of the more vocal critics will undoubtedly have offended European sensibilities. In the realm of international affairs Americans are frequently seen as being frank and out-spoken whereas Europeans are perceived as more subtle. These perceptions undoubtedly extend into the world of science and the smaller world of Shroud research. There is a need for détente and understanding among Shroud researchers from both sides of the Atlantic.

Science and religion do, in fact, have one great quality in common. They are both means to discovering truths. Science seeks to find physical truths; religion seeks to find spiritual truths. There is no reason at all why Shroud research cannot encompass both of these objectives. Past scientific studies, including STURP and the carbon-dating of the Shroud, have placed emphasis on the search for physical truths. The work of the Conservation Commission and the restoration of 2002 seem to have been focused on the preservation of the Shroud as a religious icon. In future work, neither science nor religion should take priority over the other. A way must be found to combine the needs and objectives of both. The Shroud is both a valuable scientific artifact and a unique religious relic.

One way forward could be the establishment of an international Commission on Scientific Study and Conservation of the Shroud, involving reputable scientists and researchers from both sides of the Atlantic. Such an international Commission would have to work under the guidance of the Church authorities in Turin and ultimately Rome – the Shroud is, of course, the property of the Pope. This Commission would be responsible for considering, reviewing and, where necessary, recommending for approval proposals for further studies on the Shroud. Its prime responsibility would be to give direction to future Shroud research as well as to preservation of the Shroud.

Such an idea was suggested in a petition that was drawn up in 2003 for submission to Pope John Paul II. This petition attracted 52 signatories and was posted to the Vatican. No acknowledgement of it was ever given[27]. While the proposal for a commission undoubtedly had value, the means of submitting it to the Church was totally counter-productive. The Catholic Church does not react to petitions, posted or otherwise, or to any other form of public pressure. Telling the Pope what he should do is not going to get a result. Any proposal for further co-operation on study of the Shroud will require the involvement and acceptance of the Archbishop of Turin, who is, after all, the Custodian of the Shroud on behalf of the Pope.

Prof Meacham has been particularly insistent that a new carbon-dating of the Shroud must be carried out, to either confirm or rebut the results of the 1988 tests. It was his view that this could possibly be done using material removed from the Shroud during the restoration. In pursuit of this objective he wrote a letter, together with Dr Rogers, in August 2003 to the Archbishop of Turin, Cardinal Poletti, requesting the release of material for this purpose. Cardinal Poletti responded very quickly to say that he could not grant the request. He added that:

"...a process of study and valuation has started for an organic programme of sindonological research. All the requests that the scientists presented in 2000 (but we accepted also those that arrived up to the end of 2002) are under study by a jury of scientists, from whom we await a judgement. When we have received and evaluated it, I will prepare a report to the Pope and we will do what he says."[28]

Another characteristic of the Catholic Church is that it is never in a hurry. Patience is required in dealings with the Church. Some five years have passed and there has been no outcome to the deliberations referred to by Cardinal Poletti. The material removed from the Shroud during the restoration is still stored away in Turin. That is not to say that the Church has buried the matter permanently. There is a continuing need for the international Shroud research community to engage with Turin and to press gently for progress. The establishment of the international commission suggested above could be part of that progress. Its first task, if it is established, would be to continue and complete the review of the scientific proposals submitted to the Archdiocese of Turin in 2000 and subsequently.

* * * * * * * * *

What further scientific research should be done? Prof Meacham is undoubtedly correct in proposing a new carbon-dating of the Shroud. The 1988 dating has caused so much controversy that it must be laid to rest once and for all. At present it is a spectre haunting research on the Shroud. If further tests, carried out under strict scientific procedures, conditions and supervision, confirm the 1988 results, then it will be necessary to provide a valid scientific explanation of how such a result can be reconciled with other evidence of the Shroud's age. If they provide an earlier date for the Shroud then the door will be wide open for further scientific and historical research.

It must be said that if the Shroud is eventually dated conclusively to the Middle Ages, this does not in any way invalidate the religious beliefs of the Catholic Church. It would merely prove that the Shroud is not the burial cloth of Jesus Christ. The Church has nothing to fear from further scientific research and it would be totally incorrect to suggest that its slowness in deliberating on future research work is driven by fear of the results. As the Church has already pointed out, regardless of scientific outcomes the Shroud remains a unique representation of the sufferings of Jesus Christ.

Further studies are also required on the image itself and how it was formed. Despite the efforts of STURP in 1978 and the studies of many scientists since then, how the image was formed remains a scientific mystery. Perhaps it always will be, but it continues to merit further study using the latest non-obtrusive methods known to science.

Studies that shed further light on the history of the Shroud should also be carried out. An example of such a study involves the fold-marks on the Shroud. The STURP team obtained raking light photographs of the Shroud which showed evidence of its having been folded in eight layers. These folds indicate that the Shroud may once have been kept folded in such a way that the face panel alone appeared to the viewer. This may provide evidence linking the face of Christ as described in Edessa prior to the 10th century with the folded Shroud of Turin[29]. Concern has been expressed that work done during the restoration may have damaged or destroyed these folds. It is to be hoped that this is not the case; further study of the folds alone could reveal more knowledge of the Shroud's history.

An interesting new direction of research was proposed in December 2007 – that pollen samples taken from the Shroud should be subjected to carbon-dating. Because of their high carbon content, individual pollen grains can today be dated accurately by the accelerator mass spectrometry (AMS) method of carbon-dating. The pollen samples taken from the Shroud by Dr Max Frei are in private hands in the United States of America. Therefore it may be assumed that there is no need to obtain permission from Church authorities to carry out tests on them. One of the major criticisms of the 1988 carbon-dating of the Shroud was that the samples tested all came from one location on the Shroud. Dr Frei took pollen samples from 39 different sites on the Shroud.

Indeed, carbon-dating the different species of pollen found on the Shroud could possibly indicate not only the age of the Shroud, but also shed light on the hypotheses regarding its journey from Jerusalem to Edessa, Constantinople and France. If the different pollens from different geographical locations found on the Shroud are shown to have different ages, a progression that is both chronological and geographical may be seen[30].

At the same time, further studies on the preservation of the Shroud need to be carried out. For example, the linen cloth is aging and this needs somehow to be slowed down to maintain the contrast between the body image and the background[31].

Whatever is decided upon, the approach should be multi-disciplinary, directed at both science and preservation of the Shroud, and actively involve all the parties interested in the Shroud – scientists, historians and the Church authorities.

PART 5 –
THE CHALLENGE OF THE SHROUD

Chapter 20 – Heroes, Sceptics and Challenges

In this story there are no villains, but there are some heroes. The earliest hero who can be identified was the Byzantine general John Cuercas, who led an army hundreds of miles into hostile territory simply to obtain a relic for his Emperor. He brought the Cloth of Edessa to Europe. Geoffrey de Charny would undoubtedly have done the same had he been ordered to. *Chevalier sans peur*, standard-bearer of France, his name is indelibly associated with the introduction of the Shroud as we know it to history.

Philip Lambert, counsellor to the Duke of Savoy in 1532, assisted in rescuing the Shroud from fire in the Sainte Chapelle in Chambery that year, thereby preventing its destruction. But perhaps the true hero was the blacksmith, Guillaume Pussod, who had to prise apart the grille protecting the Shroud, to get it out, because some of the officials with the various keys were not immediately available Then come some heroines, Mother Louise de Vargin and three other nuns of the Convent of the Poor Clares in Chambery, who carried out repairs in 1534 to the fire-damaged Shroud.

The agnostic French professor, Yves Delage, together with his colleague Paul Vignon, carried out the first scientific examination of photographs of the Shroud and faced the criticism of the French Academy of Sciences at the opening of the twentieth century. For Paul Vignon, study of the Shroud became almost a lifetime's work.

Other scientists too have found that what started as mere scientific curiosity led to a lifetime commitment, notably Dr John Jackson, Dr Ray Rogers and Prof Alan Adler of STURP.

Another true hero, in the mould of Guillaume Pussod, was the Turinese fireman Mario Trematore who carried the Shroud to safety when it was again threatened by fire, this time in Turin Cathedral, on an April night in 1997.

There are many other heroes in the story – some well known, some not so well known. All of them have contributed either in some way to the fact that the Shroud remains in existence today or to our knowledge of the Shroud. Secondo Pio, Fr Peter Rinaldi, Prof Pierluigi Baima-Bollone, Dr Max Frei and others cannot be forgotten.

However, overshadowing all of these is the Man on the Shroud himself. Imposing, serene, unforgettable, he is a constant presence in this story. The debate continues as to his true identity – Jacques de Molay, Leonardo da Vinci, an anonymous Crusader, an anonymous Jew, perhaps simply the creation of an anonymous genius, or Jesus Christ himself.

To the Christian believer there is no doubt, the Shroud of Turin is the shroud of Christ, risen from the dead. The believer has no need of scientific proofs, his faith is sufficient. It is not so easy for the scientists, because they seek facts and conclusions. Yet after more than a century of scientific endeavour there is still no clear answer to the two questions: how old is the Shroud and how was the image formed? The hardest task falls to the sceptics, because it is on them that the burden of proof falls. The believer needs no proof; the scientist can continue indefinitely with his search for proof; but the sceptic can only ultimately justify his scepticism by proving conclusively that the Shroud of Turin is a forgery, an artifact or indeed anything but the shroud of Christ.

Tribute must also be paid to the sceptics, because without them there would be no debate and no controversy, indeed no need to write books about the Shroud. The first sceptic was Bishop Pierre d'Arcis, who set the ball rolling in the late fourteenth century by describing the Shroud as having been "cunningly painted". Much later came the members of the French Academy of Sciences who condemned Prof Delage as a traitor to his agnosticism.

Canon Ulysse Chevalier and Fr Herbert Thurston tried to use Bishop d'Arcis's Memorandum in the early twentieth century, to the extent of selectively editing it, in pursuit of their own agenda of ridding the Catholic Church of superstition.

Science has of course produced its own sceptics. The two best known are Dr Walter McCrone and Prof Edward Hall, the latter having forcefully declared anybody who continued to believe that the Shroud was genuine, following the results of the 1988 carbon-dating, as a "Flat Earther" and "on to a loser". There are undoubtedly many other scientists who have or have had severe doubts about the Shroud, as indeed did many members of STURP at one time.

The great weakness in the argument of the sceptics is that, no matter what evidence they produce to support their case, still nobody can answer the question of how the image was produced and nobody has ever been able to duplicate it. Painting, hot statue, mediaeval photograph, powder rubbing, every proposed method has failed the tests of science and failed to adequately explain the image.

This then, is the great challenge that the Shroud presents – if it is a man-made artifact, how was it made and can it be reproduced? Without these answers the sceptics have no case.

It offers many other challenges as well, to the Catholic Church, to science, to both the faithful and the faithless and to humanity in general.

Legal ownership of the Shroud is vested in the Pope, but the Church has an obligation to all of humanity to ensure that the Shroud is properly preserved and studied. The message and teachings of Jesus Christ are not just of interest to those who count themselves his followers. They are as relevant to the world we live in today as they were in his time, with their themes of love for one's neighbour, peace, humility and tolerance. Any item with a direct association with Jesus of Nazareth is a legacy to the whole of mankind; it can never be the sole property of one group of people. Let the Catholic Church be its custodian, but never let it forget its obligations.

One challenge facing the Church is how to co-operate with science in learning more about the Shroud itself. We have seen the controversy over the restoration of 2002 and the mistrust it has caused between the scientific and religious communities. Such controversy should never be allowed to arise again. Science has a major role to play in seeking to discover physical truths about the Shroud. This search for truth has already thrown up a multitude of challenges. The STURP researchers thought that they would find the answers easily; thirty years later the answers are more elusive than ever. Scientists cannot allow themselves to be blinded by their own beliefs or pre-conceptions. Prof Yves Delage set the example that requires to be followed.

To many of the faithful, the Shroud is the shroud of Jesus, showing evidence of his resurrection from the dead. But this cannot be stated with any certainty. The Catholic Church has never stated unequivocally that it is the shroud of Christ, even if various popes have given support to this belief. Pope John Paul II referred to it merely as a "mysterious object". Although the Church itself continues to uphold the tradition of veneration of relics, to many Christians today such veneration is nothing more than superstition. Yet for Christians of a meditative frame of mind a study of the Shroud offers great scope for meditation on the Passion, death and Resurrection of Jesus. These are the core events of the Christian belief in the redemption of mankind. It is a challenge to Christians everywhere to reflect more on these.

To those who deny religious belief, or even those who simply don't bother to think very much about such things, the Shroud poses frightening questions. Is it the Shroud of Jesus Christ? Is it evidence of his resurrection from the dead? It is hard to approach such questions with intellectual dispassion if positive answers are going to challenge your entire belief system. This surely explains why so many of those who profess scepticism about the Shroud have been so vehement in their scepticism. This is a major challenge for the sceptics – the other side of the coin is that to prove that the Shroud is a mediaeval forgery would have no impact at all on the religious beliefs of Christians. Faith and knowledge are very different.

The Shroud is a legacy to all of humanity. Jesus of Nazareth is generally accepted to be a historical figure. He is mentioned, even if only in passing, by contemporary historians such as Josephus and Tacitus. There is reason to believe the Gospel of St Luke in particular to be a fair account of his public life, teachings and death. Luke was the only non-Jewish gospel writer and he seems, both in his Gospel and in the Acts of the Apostles, to have adopted the style of a classical Greek historian. Jesus was the founder and inspiration of the largest religious grouping in the world today, making up approximately one-third of mankind. To Christians and non-Christians alike, anything which enables us to better understand such a seminal figure in the history of the world must have value. Unfortunately, opportunities for understanding also stir opportunities for hatred. There are always those who wish to impose their own views and their own intolerances on the world, possibly more so today than for many years. The Shroud could be a legitimate target for destruction by fanatics and indeed one reason for the secrecy sourrounding the restoration of 2002 was to prevent against any attack on the Shroud while it was vulnerable. Protection of the Shroud against fanaticism of any type is another challenge to humanity.

* * * * * * * * *

This book has sought to examine the history of the Shroud, both speculated and known, and to put its history into the broader context of the history surrounding it. The actions of men such as the Byzantine Emperors, Otto de la Roche, Geoffrey de Charny and the Dukes of Savoy can only be fully understood in the context of the history of their times.

It has also sought to relate the evidence of the Shroud to what we know about the Passion and death of Jesus Christ. After all, the Shroud has historically been associated with Jesus and there is no other known case of a man suffering the specific injuries displayed so graphically by the man on the Shroud.

Finally, this book has sought to review the scientific studies of the Shroud, the hypotheses surrounding these studies and the results obtained, and also to look ahead at the future for research and conservation of the Shroud.

What this book emphatically does not do is to state categorically that the Shroud is the shroud of Jesus Christ. This can never be scientifically proven; in the words of Prof Alan Adler, "There is no test for Christness." What it seeks to do is to assemble evidence from over the years and present this evidence to the reader so that the reader can make up his or her own mind on the matter. The speculated history of the Shroud prior to the mid-fourteenth century can be dismissed if necessary – there is no proof that the events described involved what today is known as the Shroud of Turin. Yes, I personally believe that the Shroud is genuine, as I said right at the beginning of this book, and my views may be apparent in the style and content of the book, but it is not up to me to force my beliefs on others. Matters of faith are matters of individual choice.

The words of the American writer John Walsh, written in 1963, remain as relevant today as they were then. He wrote that the Shroud is either "one of the most ingenious, most unbelievable products of the human mind and hand on record" or it is "the most awesome and instructive relic of Jesus Christ in existence".

Which is it?

NOTES

Introduction

1. I Wilson, *The Blood and the Shroud*, (London 1999), pp 256 – 266.
2. Ibid, p 269.
3. Ibid, p 275.

Chapter 1

1. I Wilson, *Holy Faces, Secret Places*, (New York 1991), p 181.
2. I Wilson, *The Turin Shroud*, (London 1979), p 26.
3. Ibid, pp 26 – 27.
4. Ibid, p 28.
5. Ibid, p 29.
6. J Markwardt, *The Fire and the Portrait*, a paper published in 1998.
7. I Wilson, *The Turin Shroud*, (London 1979), p 30.
8. Dr L A Garza-Valdes, *The DNA of God?*, (New York 2001), p 184.
9. J G Marino and M S Benford, *Evidence for the Skewing of the C-14 Dating of the Shroud of Turin due to Repairs*, a paper published in 2000. This was followed up by a further paper by the same authors published in 2002, *Textile Evidence Supports Skewed Radiocarbon Date of Shroud of Turin*, responding to criticisms of their earlier paper.
10. I Wilson, *The Blood and the Shroud*, (London 1999), p 37.
11. I Wilson, *The Turin Shroud*, (London 1979), p 31 – 33.
12. Ibid, pp 34 – 35.
13. I Wilson, *The Blood and the Shroud*, (London 1999), pp 36 – 39.
14. Dr L A Garza-Valdes, *The DNA of God?*, (New York 2001), p 148.
15. K E Stevenson and G R Habermas, *Verdict on the Shroud*, (Pennsylvania 1982), pp 76 – 79.

16. www.shroud.com/78exam.htm.
17. www.shroud.com/78conclu.htm.
18. Dr L A Garza-Valdes, *The DNA of God?*, (New York 2001), pp 182 – 183.
19. Dr Garza-Valdes' research and findings are detailed in his book *The DNA of God?*, published in New York in 2001.
20. J G Marino and M S Benford, *Historical Support of a 16th Century Restoration in the Shroud C-14 Sample Area*, a paper published in 2002. This paper argues that sixteenth century artisans would have had the ability to insert a patch in the Shroud in such a way that it could not be seen.
21. Dr R N Rogers, *Studies on the Radiocarbon Sample from the Shroud of Turin*, published in Thermochimica Acta 425 (2005), pp 189 – 194.
22. BBC News report dated 26 January 2005. *www.news.bbc.co.uk/go/pr/fr/-/2/hi/science/nature/4210369.htm.*
23. Dr L A Garza-Valdes, *The DNA of God?*, (New York 2001), p 42.
24. Ibid, p 39.
25. Ibid, p 183.
26. I Wilson, *The New, Restored Turin Shroud*, a commentary published on the website *www.shroud.com*.
27. I Wilson, *The Blood and the Shroud*, (London 1999), pp 242 – 251.
28. C Knight and R Lomas, *The Second Messiah*, (London 1998), pp 220 – 225.
29. I Wilson, *The Blood and the Shroud*, (London 1999), pp 239 – 241.
30. I Wilson, *The Turin Shroud*, (London 1979), Appendix B, p 306.

Chapter 2

1. M Guscin, *The Sermon of Gregory Referendarius*, a paper published in 2004.
2. Ibid, Note 10.
3. *The Teaching of Addaeus the Apostle*, Vol VIII of the Ante-Nicene Fathers (American edition 1986), pp 657 – 665.

4. Eusebius Pamphili, Book 1, Chapter 13 of *Ecclesiastical History*, translated by Roy J Deferrari, Fathers of the Church Series (New York 1953), pp 76 – 82.
5. Ibid, p 77.
6. *The Teaching of Addaeus the Apostle*, Vol VIII of the Ante-Nicene Fathers (American edition 1986), editor's note 7, p 659.
7. *The Teaching of Simon Cephas*, Vol VIII of the Ante-Nicene Fathers (American edition 1986), editor's note 5, p 665.
8. *The Teaching of the Apostles*, Vol VIII of the Ante-Nicene Fathers (American edition 1986), p 670.
9. Ibid, p 671.
10. *The Acts of Sharbil*, Vol VIII of the Ante-Nicene Fathers (American edition 1986), editor's note 6, p 676.
11. S Runciman, *Byzantine Style and Civilisation*, (Middlesex, England 1975)
12. I Wilson, *The Blood and the Shroud*, (London 1999), pp 189 – 191. The Ancient Syriac Documents published in Vol VIII of the Ante-Nicene Fathers, including *The Teaching of Addaeus the Apostle*, are taken from manuscripts acquired during the nineteenth century by the British Museum from the Nitrean Monastery in Lower Egypt. The anachronisms referred to by Ian Wilson are not present in these documents, with one exception. There is one paragraph in *The Teaching of Addaeus the Apostle* (p 665) which makes anachronistic references to Bishop Serapion and Pope Zephyrinus. However, in an editorial note (p 665, note 3) the editor refers to this paragraph as "a barefaced interpolation made by some ignorant person much later."
13. J C Cruz, *Relics*, (Huntingdon, Indiana 1984), pp 1 – 8.
14. 2 Kings 13:20 – 21. All quotations from the Old and New Testaments are taken from *The Jerusalem Bible*, published in London in 1966.
15. J Markwardt, *Antioch and the Shroud*, a paper published in 1998 and revised in 1999. It needs to be emphasized again and again that there is, in fact, no documentary evidence for the Shroud ever having been in Antioch or indeed for its existence anywhere before its reported appearance in Edessa in the sixth century. Mr Markwardt's Antioch hypothesis is just

that – a hypothesis based on speculation. It must also be said, however, that it is a much likelier explanation of the Shroud's whereabouts during that period than the Abgar legend.
16. John 20:1 – 18.
17. Acts 11:18 – 21.
18. Acts 11:21.
19. J Markwardt, *Antioch and the Shroud*.
20. C P Thiede and M d'Ancona, *The Quest for the True Cross*, (London 2000), p 77. The authors describe how Symeon would have had the responsibility of orchestrating the location and identification of the True Cross and that he probably received the *Titulus Crucis*, another major Passion relic, as a precious family heirloom.
21. Ibid, p 76.
22. Eusebius Pamphili, Book 4, Chapters 5 - 6 of *Ecclesiastical History*, translated by Roy J Deferrari, Fathers of the Church Series (New York 1953). Particular reference to the persecution of the Jews in Palestine is made in Note 1 to Chapter 6, p 213.
23. K Baus, *Handbook of Church History, Vol 1 – From the Apostolic Community to Constantine*, (New York 1965), p 154.
24. Eusebius Pamphili, Book 4, Chapter 5 of *Ecclesiastical History*, translated by Roy J Deferrari, Fathers of the Church Series (New York 1953), Note 6 on p 212.
25. Ibid, Book 3, Chapter 20.
26. J Markwardt, *Antioch and the Shroud*, notes 54, 55 and 56.
27. Ibid, p 7.
28. Ibid, p 8.
29. I Wilson, *The Turin Shroud*, (London 1979), pp 112 – 113.
30. D Talbot Rice, *The Beginnings of Christian Art*, (New York 1957), Plate 8.
31. Ibid, p 66.
32. Ibid, Plate 7.
33. J Markwardt, *Antioch and the Shroud*, note 97.
34. Fr Heinrich Pfeiffer SJ, *The Shroud of Turin and the Face of Christ in Paleochristian, Byzantine and Western Mediaeval Art, Part I*, a paper published in Shroud Spectrum International no 9, Dec 1983, p 13.
35. Ibid, p 15.

36. Fr Heinrich Pfeiffer SJ, *The Shroud of Turin and the Face of Christ in Paleochristian, Byzantine and Western Mediaeval Art, Part II,* a paper published in Shroud Spectrum International no 10, Mar 1984, pp 13, 18.
37. Ibid, p 9. See also I Wilson, *The Blood and the Shroud,* (London 1999), p 189.
38. J Markwardt, *Antioch and the Shroud,* note 106.
39. Ibid, pp 11 – 12.
40. I Wilson, *The Blood and the Shroud,* (London 1999), p 312.
41. I Wilson, *The Turin Shroud,* (London 1979), pp 155 – 156.
42. M Guscin, *The Sermon of Gregory Referendarius,* a paper published in 2004, paragraphs #5 - #10.
43. J Markwardt, *Antioch and the Shroud,* notes 141 – 147.
44. J Markwardt, *The Fire and the Portrait,* a paper published in 1998, note 22. The *Acts of Thaddaeus* referred to here would seem to be the same document as *The Teaching of Addaeus the Apostle* referred to in notes 3 and 6 above, but a different translation and possibly a later text. Note 37 of Jack Markwardt's paper is also of interest.
45. A full translation of this Official History has been published as Appendix C of Ian Wilson's book *The Turin Shroud,* (London 1979), pp 313 – 331.
46. I Wilson, *The Blood and the Shroud,* (London 1999), p 176.
47. J Markwardt, *The Fire and the Portrait,* a paper published in 1998.
48. J Kilmon, *The Shroud of Turin – Genuine Article or Manufactured Relic,* a paper published in three parts in *The Glyph,* journal of the Archaeological Institute of America, San Diego, Vol 1, Nos 10 – 12 (Sept 1997, Dec 1997 and March 1998).
49. I Wilson, *The Turin Shroud,* (London 1979), p 159.
50. Fr Heinrich Pfeiffer SJ, *The Shroud of Turin and the Face of Christ in Paleochristian, Byzantine and Western Mediaeval Art, Part II,* a paper published in Shroud Spectrum International no 10, Mar 1984, p 18.
51. I Wilson, *The Blood and the Shroud,* (London 1999), pp 312 – 313. See also M Antonacci, *The Resurrection of the Shroud,* (New York 2000), pp 128 – 129, which refers to research carried out by Dr Alan Whanger of Duke University on the similarities between the two images.

52. St John of Damascus – Writings, published in the series *The Fathers of the Church* (New York 1958) and translated by F H Chase, Jr. This extract is taken from Book 4, Chapter 16, pp 372 – 373 of the volume.
53. J C Iannone, *The Mystery of the Shroud of Turin*, (New York 1998), p 110.
54. I Wilson, *The Blood and the Shroud*, (London 1999), p 176.
55. J C Iannone, *The Mystery of the Shroud of Turin*, (New York 1998), p 110.
56. I Wilson, *The Blood and the Shroud*, (London 1999), p 313.

Chapter 3

1. C Mango, *Byzantium – the Empire of New Rome*, (New York 1980), pp 16 – 17. There are numerous histories of the Byzantine Empire. In addition to Cyril Mango's relatively recent work two other useful references are N H Baynes and H St L B Moss, *Byzantium,* (Oxford University Press 1949), of which page 2 has been referred to, and A Vasiliev, *History of the Byzantine Empire,* (University of Wisconsin Press 1952), of which pages 300 – 303 have been referred to. This extract from Vasiliev's work can also be found on the webpage *www.historyofmacedonia.org/RomanMacedonia/MacedonianEpoch.htm.*
2. J C Iannone, *The Mystery of the Shroud of Turin*, (New York 1998), p 111.
3. The history of the Iconoclastic period in the Byzantine Empire is given in some detail in the entry "Iconoclasm" in the Catholic Encyclopedia, which can be found at *www.newadvent.org/cathen.*
4. I Wilson, *The Turin Shroud*, (London 1979), p 169.
5. S Runciman, *A History of the Crusades, Vol II,* (Penguin Books, London 1978), p 240.
6. R de Clari, *The Conquest of Constantinople,* translated by Edgar Holmes McNeal (Columbia Univ. Press, New York 1936), pp 103 - 104.
7. Quoted in I Wilson, *The Blood and the Shroud,* (London 1999), p 173.
8. I Wilson, *The Turin Shroud*, (London 1979), p 133.

9. Ibid, p 321.
10. Ibid, photographic plates. Similar diagrams are also shown in his later book, *The Blood and the Shroud*, (London 1999), p 177.
11. R de Clari, *The Conquest of Constantinople*, translated by Edgar Holmes McNeal (Columbia Univ. Press, New York 1936), p 112. Ian Wilson, in *The Turin Shroud*, note 37 to Chapter 18, suggests that the word "features" in this passage is better and more meaningfully rendered as "figure".
12. Ibid, p 104.
13. P Stephenson, *Relics in Constantinople*, a paper published in 2004 and available on the webpage *www.homepage.mac.com/paulstephenson/madison/byzantium/notes/RelicsinConstantinople.html*.
14. Abstract of a paper entitled *The Mandylion in Mediaeval Georgian Literature and Art*, presented by Zaza Skhirtladze at the 16th Annual Byzantine Studies Conference held in Baltimore, Maryland, USA from 26 – 28 October 1990, and referred to on the webpage *www.byzconf.org/1990abstracts.html*.
15. Z Aleksidze, *The Mandylion and the Keramion in the Ancient Literature of Georgia*, a paper published in Academia, the Journal of Human Sciences, 2001, Vol 1, Issue 1, pp 9 – 15, referred to on the webpage *www.amsi.ge/academia/0101/020101.html*.
16. I Wilson, *The Blood and the Shroud*, (London 1999), p 175.
17. Ibid, p 176.
18. M Guscin, *The Sermon of Gregory Referendarius*, a paper published in 2004. The reference is in paragraph 22 of the text of the sermon.
19. N Currer-Briggs, *The Shroud and the Grail*, (London 1987), p 53.
20. I Wilson, *The Turin Shroud*, (London 1979), p 122.
21. N Currer-Briggs, *The Shroud and the Grail*, (London 1987), p 54.
22. I Wilson, *Holy Faces, Secret Places*, (New York 1991), pp 103 – 111.
23. Ibid, p 185.
24. I Wilson, *The Blood and the Shroud*, (London 1999), pp 168 – 171.

25. D Scavone, *Greek Epitaphioi and Other Evidence for the Shroud in Constantinople up to 1204*, a paper available on the website www.shroud.com.
26. M Antonacci, *The Resurrection of the Shroud*, (New York 2000), p 144.
27. D Scavone, *Greek Epitaphioi and Other Evidence for the Shroud in Constantinople up to 1204*, a paper available on the website www.shroud.com.
28. Ibid.
29. N Currer-Briggs, *The Shroud and the Grail*, (London 1987), p 185.
30. D Scavone, *Greek Epitaphioi and Other Evidence for the Shroud in Constantinople up to 1204*, a paper available on the website www.shroud.com.
31. N Currer-Briggs, *The Shroud and the Grail*, (London 1987), pp 193 – 194.
32. D Scavone, *Greek Epitaphioi and Other Evidence for the Shroud in Constantinople up to 1204*, a paper available on the website www.shroud.com.
33. M Antonacci, *The Resurrection of the Shroud*, (New York 2000), p 145.
34. D Crispino, *Louis I, Duke of Savoy*, a paper published in Shroud Spectrum International No 7, June 1983.
35. J C Iannone, *The Mystery of the Shroud of Turin*, (New York 1998), pp 120 –121.
36. A Sinclair, *The Discovery of the Grail*, (London 1999), p 47.
37. For example, Andrew Sinclair in *The Discovery of the Grail*, (London 1999) suggests on p 200 that "… the later popes of Rome… eventually sent out their Crusaders to take Constantinople." In his book *The Shroud and the Grail*, (London 1987), Noel Currer-Briggs suggests that the cult of relics was part of the motivation for the Fourth Crusade and he develops an intricate theory that the noblemen who initiated this Crusade had the deliberate intention of seeking and capturing the burial cloths of Christ, among other relics. (pp 126ff). This strikes me as being most unlikely, as the original destination of the Fourth Crusade turned out to be Egypt and its diversion to Constantinople came about as a result of events that could not have been foreseen four years in advance.

Currer-Briggs goes so far as to describe this group as "the Shroud Mafia" (p 136). I have a strong aversion to conspiracy theories and this is one that I just cannot accept. My approach to history is based more on Murphy's Law – that if things can go wrong, they usually will - and that history is a record of excessive human frailty rather than excessive human ingenuity and foresight. This has also been described as the "cock-up" theory of history.

38. S Runciman, *A History of the Crusades, Vol III*, (Penguin Books, London 1978), pp 107 – 108.
39. Ibid, pp 108 – 109.
40. Ibid, pp 112 – 117.
41. Ibid, pp 118 – 120.
42. Ibid, pp 121 – 123.

Chapter 4

1. J Sumption, *The Hundred Years' War – Vol II: Trial By Fire*, (London 1999), p 93.
2. British Society for the Turin Shroud – Issue #45. Book review of R W Kaeuper and E Kennedy, *The Book of Chivalry of Geoffroi de Charny: Text, Context and Translation*, (Philadelphia 1996). *www.shroud.com/bsts4509.htm*.
3. J Sumption, *The Hundred Years' War – Vol II: Trial By Fire*, (London 1999), pp 61 – 62.
4. Ibid, pp 92 – 93.
5. Ibid, p 125.
6. Ibid, p 12.
7. Ibid, p 247.
8. I Wilson, *The Blood and the Shroud*, (London 1999), p 325.
9. British Society for the Turin Shroud – Issue #46. Letter to the Editor, *Was Geoffrey de Charny ever a Prisoner in England in 1342?* Pere A M Dubarle. *www.shroud.com/bsts4610.htm*.
10. D Scavone, *Geoffroy I de Charny Did Not Obtain the Present Turin Shroud on the Smyrna Campaign of 1346*, a paper published in 2005. This gives an account of de Charny's involvement in the Smyrna campaign and explains in detail why it is most unlikely that he obtained the Shroud at this time. For de Charny's presence at Aiguillon see also J Sumption, *The*

Hundred Years War – Vol I: Trial by Battle, (London 1992), p 485. Prof Scavone suggests that de Charny was knighted after the siege of Aiguillon in 1346, not in 1343. However the letter to the BSTS Newsletter referred to in Note 9 above refers to a letter dated October 1343 which makes mention of "Geoffrey de Charniz, knight, lately taken prisoner in Brittany". It also refers to a letter from King Philip VI to de Charny dated June 1343 in which he is addressed as 'knight'.

11. I Wilson, *The Blood and the Shroud,* (London 1999), pp 154 – 158.
12. Ibid, pp 324 – 325.
13. British Society for the Turin Shroud – Issue #46. Letter to the Editor, *Was Geoffrey de Charny ever a Prisoner in England in 1342?* Pere A M Dubarle. *www.shroud.com/bsts4610.htm.*
14. I Wilson, *The Blood and the Shroud,* (London 1999), pp 325 – 328.
15. G de Charny, *The Book of Chivalry,* translated by R W Kaeuper and E Kennedy, (University of Philadelphia Press 1996), p 199.
16. J Sumption, *The Hundred Years War – Vol I: Trial by Battle,* (London 1992), p 327.
17. Ibid, p 530.
18. P Ziegler, *The Black Death,* (The Folio Society, London 1997), pp 14 – 15.
19. Ibid, Chapter 4.
20. Ibid, p 48.
21. I Wilson, *The Blood and the Shroud,* (London 1999), pp 147 – 148.
22. Ibid, p 148.
23. I Wilson, *Holy Faces, Secret Places,* (New York 1991), pp 19 – 20
24. Ibid, p 61.
25. J Sumption, *The Hundred Years War – Vol II: Trial by Fire,* (London 1999), pp 160 – 161.
26. I Wilson, *The Blood and the Shroud,* (London 1999), pp 329 – 330.
27. Ibid, p 330.
28. The events in France immediately following the Battle of Poitiers are described in detail by Jonathan Sumption in

Chapters VII and VIII of *The Hundred Years War – Vol II: Trial by Fire*, (London 1999).
29. Ibid, p 384.
30. I Wilson, *The Turin Shroud*, (London 1979), Appendix B, Memorandum of Pierre d'Arcis, Bishop of Troyes, to the Avignon Pope Clement VII, p 307.
31. K Dietz, *Paralipomena on Geoffroy de Charny. Historical Notes on Geoffroy de Charny and the Shroud*, a paper presented at the Third International Dallas Conference on the Shroud of Turin, 8 – 11 September 2005.
32. Ibid.
33. I Wilson, *The Turin Shroud*, (London 1979), Appendix B, Memorandum of Pierre d'Arcis, Bishop of Troyes, to the Avignon Pope Clement VII, p 306.

Chapter 5

1. J Sumption, *The Hundred Years War – Vol II: Trial by Fire*, (London 1999), pp 478 – 479.
2. Ibid, pp 561 – 566.
3. A Tilley, *Mediaeval France*, (New York 1964), pp 122 – 123.
4. I Wilson, *The Blood and the Shroud*, (London 1999), p 331.
5. According to an article in the BSTS Shroud Newsletter issue #66, December 2007, *The "Missing Years" of the Shroud*, by Alessandro Piana, the Shroud was kept at Monfort-en-Auxois castle, a property of the de Vergy family, from 1360 – 1389. The source for this is given as M Bergeret, *Linceul de Turin – le trou historique: 1204 – 1357,* in L'identification scientifique de l'homme de Linceul, Actes du Symposium Scientifique International du CIELT, Rome, 10 – 12 June 1993, p 347.
6. I Wilson, *The Blood and the Shroud*, (London 1999), p 331.
7. Ibid, p 138.
8. Pope Gregory XI died in 1378. Although he resided in Avignon in France, he had gone to Rome for a visit. The College of Cardinals met in Rome and, under some pressure from the Romans, elected an Italian, Urban VI, as Pope. Urban VI rapidly made himself unpopular with the French cardinals in particular. A number of the cardinals reconsidered their action and carried out another election, this time for a Frenchman,

Robert of Geneva, who took the name Clement VII. This was the start of what is now known as the Western Schism. Today Clement VII is considered an Antipope, elected in opposition to a legitimate Pope. At the time, however, he had the full support of France and its allies. He was recognized as the legitimate Pope by the King of France and therefore by his subjects such as Bishop d'Arcis and Geoffrey de Charny. For this reason I have given him the title of Pope, to avoid confusion of terminology. Clement VII was a nephew of the second husband of Jeanne de Vergy, the widow of the elder Geoffrey de Charny. This would no doubt have been one reason why the younger de Charny did not hesitate to make a direct approach to the papacy to permit further expositions of the Shroud, by-passing Bishop d'Arcis. If, as will be shown, Geoffrey had reasons to be suspicious of Bishop d'Arcis's own designs on the Shroud, it would hardly be surprising that he used a family connection such as this to circumvent the Bishop.

9. K E Stevenson and G R Habermas, *Verdict on the Shroud*, (Pennsylvania 1982), pp 131 – 132.
10. H Leynen, *Le "Memoire" de Pierre d'Arcis*, a translation into French of a paper originally published in the Flemish journal *Soudarion*, September 1993, pp 14 – 17.
11. J Markwardt, *The Conspiracy against the Shroud*, a paper published in 2001.
12. It is clear from the papers of both Hilda Leynen and Jack Markwardt, referred to in the two notes above, that neither folio as it exists is dated.
13. J Markwardt, *The Conspiracy against the Shroud*, a paper published in 2001.
14. Ibid, note 9. Chevalier's transcription appeared as "Document G" on page VII of his *Etude critique sur l'origine di Saint Suaire de Lirey-Chambery-Turin*.
15. H Leynen, *Le "Memoire" de Pierre d'Arcis*, a translation into French of a paper originally published in the Flemish journal *Soudarion*, September 1993, pp 14 – 17.
16. J Markwardt, *The Conspiracy against the Shroud*, a paper published in 2001, note 8.
17. Ibid.
18. I Wilson, *The Turin Shroud*, (London 1979), Appendix B.

19. M Antonacci, *The Resurrection of the Shroud*, (New York 2000), p 151.
20. I Wilson, *The Blood and the Shroud*, (London 1999), pp 148 – 149.
21. K E Stevenson and G R Habermas, *Verdict on the Shroud*, (Pennsylvania 1982), p 133.
22. I Wilson, *The Turin Shroud*, (London 1979), p 308.
23. I Wilson, *The Blood and the Shroud*, (London 1999), p 332 - 334.
24. I Wilson, *The Turin Shroud*, (London 1979), p 309.
25. The family trees of the Valois Kings of France and Dukes of Burgundy are shown in the *Oxford Illustrated History of Medieval Europe* (Oxford University Press 1988), pp 312 – 314. This also refers to Philip the Bold's marriage to Margaret, daughter of Count Louis of Flanders and Artois.
26. The events of the early years of the reign of Charles VI, including his madness, are described in W T Waugh, *A History of Europe 1378 – 1494,* Vol IV of Methuen's History of Mediaeval and Modern Europe (3rd Edition, London 1949), pp 17 - 30. The madness of Charles VI showed similar characteristics to that of King George III in England four centuries later. King George's madness has been attributed by some authorities to the hereditary condition porphyria. It is perhaps worth speculating as to whether porphyria was also the cause of Charles VI's madness – George III was a direct descendant of his through the second marriage of his daughter Catherine of Valois to Owen Tudor.
27. Ibid, pp 34 – 54. These pages describe the events in France from 1413 to 1420.
28. I Wilson, *The Blood and the Shroud*, (London 1999), p 334.

Chapter 6

1. Dorothy Crispino's investigation at Castle Montfort, near Montbard, is described in an unpublished document sent to the author by her. This document quotes Sanna Solaro as writing in *La Santa Sindone che si venera a Torino,* 1901. There is a photograph of the Montbard castle, taken by Dorothy

Crispino, on the front cover of Shroud Spectrum No 8, Sept 1983.
2. I Wilson, *The Blood and the Shroud*, (London 1999), p 335.
3. W T Waugh, *A History of Europe 1378 – 1494*, Vol IV of Methuen's History of Mediaeval and Modern Europe (3rd Edition, London 1949), pp 54 – 66.
4. Ibid, pp 71 – 79.
5. Ibid, pp 80 – 91.
6. *www.crimelibrary.com/criminal_mind/scams/shroud_of_turin/2.html?sect=27.*
7. I Wilson, *The Blood and the Shroud*, (London 1999), p 335.
8. Article in the British Society for the Turin Shroud Newsletter Issue #48. *www.shroud.com/bsts4805.htm.*
9. I Wilson, *The Blood and the Shroud*, (London 1999), p 335.
10. Ibid, p 336.
11. It is a curious fact that the year 1453 marks three historical but unrelated events: the fall of Constantinople to the Turks, the final capture of Bordeaux by the French and the end of the Hundred Years' War, and possibly the transfer of ownership of the Shroud by Margaret de Charny to the House of Savoy. No doubt a conspiracy theorist could make much of this.
12. The early history of the House of Savoy is taken from *The Royal House of Savoy, an Introduction,* by B I Di Bella, published on the webpage *www.regalis.com/reg/savhistory.htm.*
13. Further information on the origins and early history of the House of Savoy can be found in the Catholic Encyclopedia at *www.newadvent.org/cathen/13492a.htm,* and also at *www.nationalflaggen.de/flags-of-the-world/flags/fr-savoy.html.*
14. Count Amadeus V is shown in the family tree of Count Amadeus III of Geneva, his grandson, on the webpage *www.freepages.genealogy.rootsweb.com/~jamesdow/s030/f165482.htm.*
15. The exact relationship between Count Amadeus III of Geneva and Aymon of Geneva, the second husband of Jeanne de Vergy, is described in an unpublished document headed *Cousins and Connections,* sent to the author by Dorothy Crispino. Amadeus and Aymon shared a common grandfather in Count Amadeus II of Geneva. One of his sons became Count William III of Geneva and married Agnes of Savoy, the daughter of Count

Amadeus V "the Great" of Savoy. One of their children became Count Amadeus III of Geneva. The youngest son of Amadeus II of Geneva was Hugh, whose eldest son was Aymon, the eventual husband of Jeanne de Vergy.
16. N Currer-Briggs, *The Shroud and the Grail,* (London 1987), p 46. As the various genealogies detailed in Noel Currer-Briggs' book show, there was a great deal of inter-marriage among the leading families of Europe, resulting in a highly complex web of family relationships.
17. D Crispino, *Louis I, Duke of Savoy,* a paper published in Shroud Spectrum International No 7, June 1983.
18. Ibid.
19. Ibid, p 13.
20. The events that took place in Savoy involving the Duke, Francois de la Palud and Margaret de Charney are described in detail in *Memoires de L'Academie des Sciences Belles-Lettres et Arts de Savoie (sixieme serie), Tome IV,* published in 1960, pp 82 – 91. I owe many thanks to Dorothy Crispino for sending me these pages from this document, together with some brief notes summarizing the story.
21. I Wilson, *The Turin Shroud,* (London 1979), pp 243 – 244.
22. I Wilson, *The Blood and the Shroud,* (London 1999), pp 336 - 337.
23. Ibid, pp 337 - 338

Chapter 7

1. I Wilson, *The Turin Shroud,* (London 1979), p 243.
2. D Crispino, *Louis I, Duke of Savoy,* a paper published in Shroud Spectrum International No 7, June 1983.
3. I Wilson, *The Turin Shroud,* (London 1979), p 245.
4. I Wilson, *The Blood and the Shroud,* (London 1999), p 337.
5. Ibid, pp 337 – 339. Savigliano is a monastery south of Turin which is not to be confused with the town of Avigliano, west of Turin, which is also associated with the travels of the Shroud at this time. Avigliano is at the start of a route through the Alps connecting Turin and Chambery.
6. Ibid, p 339.

7. L Picknett and C Prince, *Turin Shroud – in whose Image?* (London 1994), p 110. They go so far as to claim in their note 8 to Chapter 6 (p 200) that they had scoured the Shroud literature and historical records for any reference to public or private showings of the Shroud during this period and had found none. However Ian Wilson in *The Blood and the Shroud* does give details of such expositions. Wilson is highly critical of the Picknett/Prince theory and sums up his view on page 243 of *The Blood and the Shroud* with the words "Intriguing idea; shame about the facts."
8. Ibid, p 68.
9. Ibid, p 178.
10. Ibid, p 112.
11. I Wilson, *The Blood and the Shroud*, (London 1999), p 244.
12. L Picknett and C Prince, *Turin Shroud – in whose Image?* (London 1994), pp 112 – 113.
13. S Bramley, *Leonardo – the Artist and the Man*, (London 1992), pp 228 – 234.
14. I Wilson, *The Blood and the Shroud*, (London 1999), p 340.
15. I Wilson, *The Turin Shroud*, (London 1979), p 246.
16. I Wilson, *The Blood and the Shroud*, (London 1999), pp 342 – 343.
17. Ibid, p 78.
18. M S Benford and J G Marino, *Evidence for the Skewing of the C-14 Dating of the Shroud of Turin due to Repairs,* a paper published in 2000, *Textile Evidence Supports Skewed Radiocarbon Date of Shroud of Turin,* a paper published in 2002, and *Historical Support of a 16th Century Restoration in the Shroud c-14 Sample Area,* a paper published in 2002. In these three papers the authors write in depth about their hypothesis that a "patch" of material from the 16th century was skillfully woven into the original Shroud cloth and that this patch contaminated the materials taken for the radiocarbon dating tests, resulting in an incorrect date for the Shroud.
19. I Wilson, *The Blood and the Shroud*, (London 1999), p 345.
20. Ibid, p 75.
21. Ibid, pp 75 – 76, 345 – 347.
22. I Wilson, *The New, Restored Turin Shroud,* a commentary published on the website *www.shroud.com*.

23. E Weber, *A Modern History of Europe,* (New York 1971), p 266.
24. E J Knapton, *Europe 1450 – 1815, Vol I,* (New York 1961), pp 87 – 88, 92.
25. N Davies *Europe – a History,* (Oxford University Press 1996), p 525.
26. E J Knapton, *Europe 1450 – 1815, Vol I,* (New York 1961), p 92.
27. E Weber, *A Modern History of Europe,* (New York 1971), p 266.
28. I Wilson, *The Blood and the Shroud,* (London 1999), p 347.
29. E Weber, *A Modern History of Europe,* (New York 1971), p 144.
30. I Wilson, *The Blood and the Shroud,* (London 1999), p 347.
31. *www.regalis.com/reg/savhistory.htm.*
32. I Wilson, *The Blood and the Shroud,* (London 1999), pp 348 – 370.
33. I Wilson, *The Blood and the Shroud,* (London 1999), pp 2 – 3.
34. I Wilson, *The New, Restored Turin Shroud,* a commentary published on the website *www.shroud.com.*
35. J Markwardt, *The Conspiracy against the Shroud,* a paper published in 2001.
36. I Wilson, *The Turin Shroud,* (London 1979), p 231. Ian Wilson believes that the Shroud came to the de Charny family from the Knights Templar, and that Clement had some knowledge of the Templars' association with the Shroud.
37. I Wilson, *The Blood and the Shroud,* (London 1999), p 142.
38. Ibid, p 334.
39. Ibid, pp 356 – 357.
40. Ibid, pp 337 – 343.
41. Dr P Barbet, *A Doctor at Calvary,* tr. The Earl of Wicklow, (New York 1953), p 19.
42. Ibid, pp 19 – 21.
43. I Wilson, *The Blood and the Shroud,* (London 1999), p 361.

Chapter 8

1. I Wilson, *The Blood and the Shroud*, (London 1999), p 8.
2. Ibid, p 9.
3. Ibid, p 213.
4. British Society for the Turin Shroud – Issue # 56. C Barta, *The Shroud sent to Louis IX of France by Baldwin II, the Latin Emperor at Constantinople*.
5. J C Cruz, *Relics*, (Huntingdon, Indiana 1984), pp 5 – 6.
6. Ibid.
7. I Wilson, *The Turin Shroud*, (London 1979), p 308.
8. J Sumption, *The Albigensian Crusade*, (London 1999), pp 32 – 34.
9. J Pelikan, *The Emergence of the Catholic Tradition (100 – 600); Volume 1 of The Christian Tradition: A History of the Development of Doctrine*, (Chicago 1975), pp 75 – 76.
10. J Sumption, *The Albigensian Crusade*, (London 1999), p 35.
11. Ibid, pp 39 – 42.
12. J Markwardt, *Was the Shroud in Languedoc during the Missing Years*, a paper published in 1997, and *The Cathar Crucifix: New Evidence of the Shroud's Missing History*, published in 2000, both available on the website *www.shroud.com*,
13. J Markwardt, *Was the Shroud in Languedoc during the Missing Years*, pp 1 – 2.
14. Ibid, note 4.
15. J Sumption, *The Albigensian Crusade*, (London 1999), pp 63 – 67.
16. Ibid, p 225.
17. Ibid, pp 237 – 240.
18. Ibid, p 243.
19. J Markwardt, *Was the Shroud in Languedoc during the Missing Years*, p 4.
20. Ibid, p 6.
21. J Markwardt, *The Cathar Crucifix: New Evidence of the Shroud's Missing History*.
22. J J Robinson, *Dungeon, Fire and Sword*, (London 1994), pp 32 – 46.
23. Ibid, pp 96 – 97.
24. Ibid, p 403.

25. Ibid, pp 418 – 420.
26. Ibid, p 425.
27. Ibid, pp 427 – 428.
28. Ibid, p 432.
29. Ibid, pp 434 – 464.
30. Ibid, pp 467 – 468.
31. K E Stevenson and G R Habermas, *Verdict on the Shroud*, (Pennsylvania 1982), p 28. I Wilson, *The Turin Shroud*, (London 1979), pp 215 – 224.
32. I Wilson, *The Turin Shroud*, (London 1979), p 216.
33. I Wilson, *The Blood and the Shroud*, (London 1999), pp 156 – 157.
34. R de Clari, *The Conquest of Constantinople*, translated by Edgar Holmes McNeal (Columbia Univ. Press, New York 1936), pp 31 – 34.
35. I Wilson, *The Turin Shroud*, (London 1979), pp 199ff.
36. I Wilson, *The Turin Shroud*, (London 1979), p 212, quoting from Heinrich Finke, *Papsttum und Untergang des Templerordens* (Munster 1907) II, p 334.
37. Ibid, p 212, quoting from *Chronicles of St Denis,* art III, quoted in de Puy, *Histoire de l'Ordre Militaire des Templiers* (1713), I, p 25. In Chapter 19 of *The Turin Shroud* Ian Wilson has clearly carried out detailed research in old documents on the subject of the mysterious head and references to it in the Templar interrogations. While there is still no proof that any such head existed, his research is sufficient to leave a question mark on this point.

Chapter 9

1. A history of the Byzantine Empire at the time of the Fourth Crusade and of the various successor states, including genealogies of leading families of the time, which include those of Theodore Angelus Ducas and Otto de la Roche, can be found on the webpages *www.friesian.outremer.htm* and *www.friesian.romania.htm*.
2. Quoted in full in N Currer-Briggs, *The Shroud and the Grail*, (London 1987), p 148.

3. D Scavone, *The Turin Shroud in Constantinople: The Documentary Evidence,* in R F Sutton, ed. *Daidalikon, Studies in Memory of Rev R V Schoder, SJ,* Wauconda, Illinois: Bolchazy-Carducci, 311 – 329; quoted at length in an e-mail from Dan Scavone to the author dated 23 May 2005.
4. C Barta, *The Shroud sent to Louis IX of France by Baldwin II, the Latin Emperor at Constantinople.* British Society for the Turin Shroud – Newsletter Issue # 56.
5. R de Clari, *The Conquest of Constantinople,* translated by Edgar Holmes McNeal (Columbia Univ. Press, New York 1936), p 112.
6. Ibid, p 112.
7. S Runciman, *A History of the Crusades, Vol III,* (Penguin Books, London 1978), p 123.
8. N Currer-Briggs, *The Shroud and the Grail,* (London 1987), p 152. Currer-Briggs says that Guy was Otto de la Roche's nephew, but an article by Alessandro Piana in the BSTS Shroud Newsletter issue #66, December 2007, *The "Missing Years" of the Shroud,* states that Guy was one of Otto's sons. This seems more feasible as it is unlikely that Otto would have left such a valuable possession as Athens to a nephew when he had a son of his own, Otto II.
9. Ibid, pp 49 – 50. Mgr Barnes actually suggested that Otto de la Roche sent it to his father, Pons de la Roche. This would not have been possible, as Pons died in 1203, before the Fourth Crusade.
10. I Wilson, *The Blood and the Shroud,* (London 1999), pp 156 – 157. Ian Wilson gives a detailed geneology of the de Charny and de Vergy families in his book. Similar geneologies showing the descent of Jeanne de Vergy from Otto de la Roche are also contained in N Currer-Briggs, *The Shroud and the Grail,* (London 1987), pp 35 and 38 – 39, and A Piana, *The "Missing Years" of the Shroud,* an article published in the BSTS Shroud Newsletter issue #66, December 2007, pp 22 – 24.
11. C Barta, *The IV CIELT International Symposium, Paris 2002,* a Report published in the Newsletter of the British Society for the Turin Shroud. *www.shroud.com/pdfs/n56part6.pdf.*

12. A Piana, *The "Missing Years" of the Shroud,* an article published in the BSTS Shroud Newsletter issue #66, December 2007, p 17.
13. Cesar Barta has described his examination of the chest in an e-mail to the author dated 28 July 2005.
14. My information on the capture of Constantinople, the relics in the Bucoleon Palace and the activities of Boniface of Montferrat and the Empress Maria is taken from Noel Currer-Brigg's book *The Shroud and the Grail,* pp 141 – 158. His account is mistaken, however, in that he places the Shroud in the Bucoleon Palace and the Mandylion, which he calls the Sudarium, in the Blachernae Palace. This is quite the opposite of what is stated by Robert de Clari. On this basis he suggests that it was the Shroud that was taken to Hungary by the Empress Maria. If one takes Robert de Clari at his word, it would have been the Mandylion that was in the Empress's possession, the Shroud having gone to Athens according to Theodore Angelus Ducas.
15. In his account of the Empress Maria, Noel Currer-Briggs goes on to suggest that she gave the Shroud to her brother King Andrew and that his successor as King of Hungary passed it to the Templars for a financial consideration (*The Shroud and the Grail,* pp 164 – 165.) Elsewhere in his book he develops another elaborate conspiracy theory to explain how the Shroud then passed from the Templars and finally came into the hands of Geoffrey de Charny (pp 101 – 117). As we have seen, if the Empress Maria had anything to pass on to her Hungarian relatives, it would have been the Mandylion, not the Shroud. The disappearance of any Templar "idol" or icon with other Templar treasure is a much simpler explanation for me than any conspiracy theory. I would also note that, if the Templars had had the Shroud rather than the Mandylion, they would undoubtedly have been aware of its full-length image, and this would have come out at some point in the interrogation of the captured Templars. The fact that that was never mentioned indicates to me that, if they had anything, the Templars only had a depiction of a head – the Mandylion.
16. A Sinclair, *The Sword and the Grail,* (London 1993), p 42.

17. E-mail to the author from Stuart Beattie, Project Director of the Rosslyn Chapel Trust, dated 25 January 2005.
18. C Knight and R Lomas, *The Second Messiah*, (London 1998), Plate 5.
19. I Wilson, *The Turin Shroud*, (London 1979), pp 207 – 209. See also I Wilson, *The Blood and the Shroud*, (London 1999), p 158.
20. According to Ian Wilson, writing in *The Blood and the Shroud*, p 184, the American Shroud researcher Professor Alan Whanger of Duke University in North Carolina has developed a special polaroid projection technique that shows up points of "congruence" between the face on the Turin Shroud and similar icons or images. Prof Whanger published details of this technique in a paper entitled *Polarised Overlay Technique; A New Image Comparison Method and Its Applications*, which appeared in Applied Optics 24, no 16, 15 March 1985, pp 766 – 72. In an e-mail to the author dated 15 July 2005, Ian Wilson said that Prof Whanger has applied this technique to the Templecombe panel, from which he claims a very close match between the Turin Shroud image and the panel. The author has carried out his own comparison using the markings defined by the French scholar Paul Vignon in 1939, which are referred to by Ian Wilson in *The Blood and the Shroud*, p 184 and plate 39a. Vignon identified a number of recurring oddities on early portraits of Christ which correspond to identical features on the Shroud. There are fifteen of these and they are referred to as "Vignon markings". I have identified nine of these "Vignon markings" on a photograph of the Templecombe panel which was sent to him by Ian Wilson. These nine are the following:
    a. 'V' shape at the bridge of the nose.
    b. Raised right eyebrow.
    c. Heavily accented, "owlish" eyes.
    d. Accentuated left cheek.
    e. Accentuated right cheek.
    f. Enlarged left nostril.
    g. Heavy line under lower lip.
    h. Hairless area between lower lip and beard.
    i. Forked beard.
21. I Wilson, *Holy Faces, Secret Places*, (New York 1991), pp 50 – 54.

22. The fact that the Templecombe panel is dated to around 1280 does not exclude the possibility that it is from a slightly later date, such as 1300. Dating of such artifacts is not that precise.
23. I Wilson, *The Blood and the Shroud,* (London 1999), pp 328 – 329.
24. N Currer-Briggs, *The Shroud and the Grail*, (London 1987), p 50.
25. D Scavone, *On Besancon and Other Plausible Theories for the Shroud during the Missing 150 years, 1204 to 1355*, a paper presented at the Shroud Science Group International Conference held at Ohio State University in August 2008.
26. N Currer-Briggs, *The Shroud and the Grail*, (London 1987), pp 50 – 51.
27. D Scavone, *On Besancon and Other Plausible Theories for the Shroud during the Missing 150 years, 1204 to 1355*, a paper presented at the Shroud Science Group International Conference held at Ohio State University in August 2008.
28. N Currer-Briggs, *The Shroud and the Grail*, (London 1987), p 62.
29. Ibid, p 51, quoting a paper by Jules Gauthier published by the Academie des Sciences, Belles-Lettres et Arts de Besancon, in 1883.
30. Extract from the webpage *www.classscoop.sunherald.com.au*.
31. A Piana, *The "Missing Years" of the Shroud,* an article published in the BSTS Shroud Newsletter issue #66, December 2007, pp 19 - 20. This article contains a photograph of the copy of the Shroud of Besancon at Ray-sur-Saone. It is very similar, in both style and detail to a photograph of a copy of the Shroud of Besancon in L Kestrel, *"Whips and Angels": Painting on Cloth in the Mediaeval Period*, a paper available at *www.shroud.com*, p 11. This photograph is in turn taken from D Wolfthal, *Beginnings of Netherlandish Canvas Painting: 1400 – 1530,* (Cambridge University Press, 1989), p 5.

Chapter 10

1. Dr F T Zugibe, *The Crucifixion of Jesus, a Forensic Enquiry,* (New York 2005) pp 51 – 54.

2. I Wilson, *The Blood and the Shroud*, (London 1999), p 51.
3. Dr F T Zugibe, *The Crucifixion of Jesus, a Forensic Enquiry*, (New York 2005) p 55.
4. I Wilson, *The Blood and the Shroud*, (London 1999), p 51.
5. Luke 23:33.
6. John 20:25.
7. Dr F T Zugibe, *The Crucifixion of Jesus, a Forensic Enquiry*, (New York 2005) pp 52 - 53.
8. Ibid, p 56.
9. Ibid, pp 40 – 41.
10. C P Thiede and M d'Ancona, *The Quest for the True Cross*, (London 2000), pp 73 – 74.
11. Dr Pierre Barbet, in his book *A Doctor at Calvary*, (New York 1953), on pages 48 and 55, suggests that it was the Roman custom for the condemned to walk naked to the place of execution, carrying the *patibulum*. He further suggests that this was not done in Jerusalem, in deference to local customs.
12. Ibid, p 66.
13. Dr F T Zugibe, *The Crucifixion of Jesus, a Forensic Enquiry*, (New York 2005) p 59.
14. Ibid, p 57.
15. Ibid, p 58. Dr Pierre Barbet also refers to this as a product of artistic invention, having been first mentioned by Gregory of Tours in the sixth century – *A Doctor at Calvary*, p 46.
16. Dr P Barbet, *A Doctor at Calvary*, tr. The Earl of Wicklow, (New York 1953), p 98.
17. Ibid, pp 101 – 105.
18. Dr F T Zugibe, *The Crucifixion of Jesus, a Forensic Enquiry*, (New York 2005) pp 72 - 75.
19. Ibid, p 77.
20. Ibid, pp 81 – 89.
21. Ibid, p 91.
22. J Markwardt, *The Cathar Crucifix: New Evidence of the Shroud's Missing History*, a paper published in 2000, note 18.
23. Ibid, note 21.
24. Ibid, note 53.
25. N Currer-Briggs, *The Shroud and the Grail*, (London 1987), pp 191 – 192.

26. Dr F T Zugibe, *The Crucifixion of Jesus, a Forensic Enquiry*, (New York 2005) p 95.
27. Dr P Barbet, *A Doctor at Calvary*, tr. The Earl of Wicklow, (New York 1953), p 48. Dr F T Zugibe, *The Crucifixion of Jesus, a Forensic Enquiry*, (New York 2005) pp 19 - 20.
28. Dr F T Zugibe, *The Crucifixion of Jesus, a Forensic Enquiry*, (New York 2005) p 20.
29. Dr P Barbet, *A Doctor at Calvary*, tr. The Earl of Wicklow, (New York 1953), pp 55 – 56.
30. Ibid, p 49.
31. Dr F T Zugibe, *The Crucifixion of Jesus, a Forensic Enquiry*, (New York 2005) p 92.
32. Ibid, p 97.
33. Dr P Barbet, *A Doctor at Calvary*, tr. The Earl of Wicklow, (New York 1953), pp 74 – 76. Also Appendix I, p 174.
34. Ibid, pp 76 – 78.
35. Dr F T Zugibe, *The Crucifixion of Jesus, a Forensic Enquiry*, (New York 2005) pp 108 - 122.
36. Ibid, pp 129 – 131.
37. Ibid, p 132.
38. Ibid, p 133.
39. Ibid, p 106.
40. Dr P Barbet, *A Doctor at Calvary*, tr. The Earl of Wicklow, (New York 1953), p 51.
41. Dr F T Zugibe, *The Crucifixion of Jesus, a Forensic Enquiry*, (New York 2005) p 144.
42. Dr F T Zugibe, *The Crucifixion of Jesus, a Forensic Enquiry*, (New York 2005) p 212.

Chapter 11

1. Luke 22:2.
2. John 13:21
3. Mark 14:32.
4. Luke 22:41 – 44.
5. Dr F T Zugibe, *The Crucifixion of Jesus, a Forensic Enquiry*, (New York 2005) p 8.
6. Ibid, p 9.
7. Mark 14:34 – 42.

8. Luke 22:51.
9. Rev R Gorman CP, *The Last Hours of Jesus*, (New York 1960), p 45.
10. Ibid, p 86. I have measured approximate distances in and around Jerusalem from a map in the front of a book by Josef Blinzler, translated by Isobel and Florence McHugh, *The Trial of Jesus*, (Newman Press, Maryland 1959).
11. Mark 14:53 – 64.
12. Luke 22:63 – 66.
13. *www.livius.org/pi-pm/pilate/pilate01.htm*.
14. J Blinzler, *The Trial of Jesus*, translated by Isobel and Florence McHugh, (Newman Press, Maryland 1959), p 177.
15. The Catholic Encyclopedia. *www.newadvent.org/cathen/12083c.htm*.
16. *www.livius.org.pi-pm/pilate/pilate04.html*.
17. *www.livius.org.pi-pm.pilate/pilate05.html*.
18. Luke 23:2.
19. Luke 23:6 – 7.
20. Rev R Gorman CP, *The Trial of Christ: A Re-Appraisal*, (Our Sunday Visitor, Indiana 1972), p 139. Fr Gorman states that this practice was not exclusively Jewish and refers to a similar incident recorded in an Egyptian papyrus dated 86 – 88 AD. In *The Trial of Jesus*, Excursus X, pp 218 – 221, Josef Blinzler refers to a Jewish Mishna tractate, Pesachim VIII 6a, which appears to imply the custom.
21. John 18:39 – 40.
22. Luke 23:14 – 16.
23. John 19:1.
24. Dr F T Zugibe, *The Crucifixion of Jesus, a Forensic Enquiry*, (New York 2005) pp 21 – 22. I am greatly indebted to Dr Zugibe in this Chapter for his clinical but graphic description of the suffering of Jesus and of the physical effects of the treatment meted out to him.
25. Ibid, p 25.
26. John 19:2 – 3.
27. Dr F T Zugibe, *The Crucifixion of Jesus, a Forensic Enquiry*, (New York 2005) pp 33 – 36.
28. Ibid, p 34, quoting Dr Robert Nugent, Professor and Chairman of the Department of Neurosurgery at West Virginia School

of Medicine, writing in *West Virginia University Newsletter*, 1986.
29. Ibid, p 36.
30. John 19:6.
31. John 19:19 – 22.
32. Luke 23:26.
33. Dr F T Zugibe, *The Crucifixion of Jesus, a Forensic Enquiry*, (New York 2005) p 100.
34. John 19:25. Only John, of the four evangelists, mentions the presence of Mary at Jesus's crucifixion. However, John claims his description to be that of an eye-witness, so he was presumably in a position to know that she was there.
35. Luke 23:35 – 39.
36. Mark 15:26, 34 – 37.
37. Dr F T Zugibe, *The Crucifixion of Jesus, a Forensic Enquiry*, (New York 2005) p 133.
38. John 19:28 – 29.
39. Dr F T Zugibe, *The Crucifixion of Jesus, a Forensic Enquiry*, (New York 2005) p 135.
40. John 19:33 – 34.
41. Dr F T Zugibe, *The Crucifixion of Jesus, a Forensic Enquiry*, (New York 2005) p 140.
42. Dr F T Zugibe, *The Crucifixion of Jesus, a Forensic Enquiry*, (New York 2005), p 189.
43. John 19:38 describes how Joseph was a secret follower of Jesus. Another secret follower, Nicodemus, was also involved in the burial of Jesus. Luke 23:52 describes how Joseph wrapped the body of Jesus in a shroud. Mark 15:46 states that the shroud was brought by Joseph. Matthew 27:60 states that the tomb was Joseph's own, that he had cut out of rock.
44. Dr F T Zugibe, *The Crucifixion of Jesus, a Forensic Enquiry*, (New York 2005) p 213.
45. John 19:40.

Chapter 12

1. I Wilson, *The Turin Shroud*, (London 1979), p 26.
2. Dr F T Zugibe, *The Crucifixion of Jesus, a Forensic Enquiry*, (New York 2005) p 185.

3. Ibid, p 190.
4. I Wilson, *The Turin Shroud*, (London 1979), p 42.
5. Dr F T Zugibe, *The Crucifixion of Jesus, a Forensic Enquiry*, (New York 2005) p 189.
6. I Wilson, *The Turin Shroud*, (London 1979), p 39.
7. Dr P Barbet, *A Doctor at Calvary*, tr. The Earl of Wicklow, (New York 1953), p 24.
8. Ibid, p 28.
9. Ibid, pp 84 – 87.
10. Ibid, Appendix II, p 176.
11. I Wilson, *The Turin Shroud*, (London 1979), pp 41 – 43.
12. Ibid, p 43, quoting unpublished notes written by Dr Willis.
13. Dr P Barbet, *A Doctor at Calvary*, tr. The Earl of Wicklow, (New York 1953), p 88.
14. Ibid, pp 88 – 90.
15. I Wilson, *The Turin Shroud*, (London 1979), p 45.
16. Dr F T Zugibe, *The Crucifixion of Jesus, a Forensic Enquiry*, (New York 2005) p 180.
17. Ibid, p 195.
18. I Wilson, *The Turin Shroud*, (London 1979), p 45.
19. Ibid, p 47.
20. Dr F T Zugibe, *The Crucifixion of Jesus, a Forensic Enquiry*, (New York 2005) p 116.
21. Ibid, p 114.
22. Ibid, p 223.
23. Ibid, p 193.
24. Dr P Barbet, *A Doctor at Calvary*, tr. The Earl of Wicklow, (New York 1953), p 104.
25. Dr F T Zugibe, *The Crucifixion of Jesus, a Forensic Enquiry*, (New York 2005), p 193.
26. Ibid, p 196.
27. M Antonacci, *The Resurrection of the Shroud*, (New York 2000), p 20.
28. Dr R Bucklin, *An Autopsy on the Man of the Shroud*, a paper published in 1997.
29. Ibid.
30. E Jumper, K Stevenson Jr and J Jackson, *Images of Coins on a Burial Cloth?*, article in The Numismatist, July 1978.

31. J C Iannone, *The Mystery of the Shroud of Turin*, (New York 1998), pp 33 – 42.
32. Dr R M Haralick, *Analysis of Digital Images of the Shroud of Turin*, (Spatial Data Analysis Laboratory, Virginia Polytechnic Institute and State University, Blacksburg, Virginia 24061, December 1983), quoted in J C Iannone, *The Mystery of the Shroud of Turin*, (New York 1998), p 40.
33. F Filas SJ, *The Dating of the Shroud of Turin from Coins of Pontius Pilate*, (Cogan Productions, a Division of ACTA Foundation, January 1984), 19, quoted in J C Iannone, *The Mystery of the Shroud of Turin*, (New York 1998), p 38.
34. J C Iannone, *The Mystery of the Shroud of Turin*, (New York 1998), p 38. This is also referred to in a letter to the Editor of the Newsletter of the British Society for the Turin Shroud, from Dr Antonio Lombatti of Pontremoli in northern Italy, published in Issue # 45 of the Newsletter.
35. R Hachlili, *Ancient Burial Customs Preserved in Jericho Hills*, Biblical Archaeology Review 5.4 (July 1979), 28 – 35, quoted in M Antonacci, *The Resurrection of the Shroud*, (New York 2000), p 106.
36. M Antonacci, *The Resurrection of the Shroud*, (New York 2000), p 106.
37. A Lombatti, *Doubts Concerning the Coins over the Eyes*, letter to the Editor of the Newsletter of the British Society for the Turin Shroud, Issue # 45.
38. M Antonacci, *The Resurrection of the Shroud*, (New York 2000), p 107.
39. Z Greenhut, *Burial Cave of the Caiaphas Family*, Biblical Archaeology Review (September/October 1992): 28 – 36, 76, quoted in M Antonacci, *The Resurrection of the Shroud*, (New York 2000), p 108.
40. A Lombatti, *Doubts Concerning the Coins over the Eyes*, letter to the Editor of the Newsletter of the British Society for the Turin Shroud, Issue # 45.
41. Quoted on the webpage *www.shroudstory.com/faq-coins.htm*. Also referred to by Dr F T Zugibe, *The Crucifixion of Jesus, a Forensic Enquiry*, (New York 2005), pp 237 - 238.
42. e-mail from Barrie Schwortz to the author dated 21 June 2006.

43. I Wilson, *The Turin Shroud*, (London 1979), pp 63 – 65.
44. M Antonacci, *The Resurrection of the Shroud*, (New York 2000), pp 117 – 118.
45. J C Iannone, *The Mystery of the Shroud of Turin*, (New York 1998), p 89.
46. Dr F T Zugibe, *The Man of the Shroud was Washed*, a paper published in Sindon N S, Quad no 1, June 1989.
47. M Antonacci, *The Resurrection of the Shroud*, (New York 2000), p 118.
48. Ibid.
49. Dr F T Zugibe, *The Crucifixion of Jesus, a Forensic Enquiry*, (New York 2005), p 227.

Chapter 13

I would particularly like to record my gratitude to Mark Guscin for sending me a manuscript copy of his book *The History of the Sudarium of Oviedo* for my reference. Mark is undoubtedly one of the leading experts on the Sudarium of Oviedo and I have drawn extensively on his work in this Chapter.

1. John 20:6 – 7.
2. M Guscin, *The History of the Sudarium of Oviedo*, manuscript copy, pp 10 – 11, 14. See also M Guscin, *The Sudarium of Oviedo: Its History and Relationship to the Shroud of Turin*, a paper published in 1997.
3. M Guscin, *Recent Historical Investigations on the Sudarium of Oviedo*, a paper published by the Investigation Team of the Centro Espanol de Sindonologia in 1999.
4. M Guscin, *The History of the Sudarium of Oviedo*, manuscript copy, p 18.
5. Ibid, p 20.
6. Ibid, p 23.
7. Ibid, p 23.
8. Ibid, p 28.
9. Ibid, pp 38 – 41.
10. Ibid, pp 53 – 56.
11. M Guscin, *The Sudarium of Oviedo: Its History and Relationship to the Shroud of Turin*, a paper published in 1997.

12. G Heras, J-D Villalain and J-M Rodriguez, *Comparative Study of the Sudarium of Oviedo and the Shroud of Turin*, translated by Mark Guscin, a paper presented at the 3rd International Congress for the Study of the Shroud of Turin, June 1998.
13. Ibid.
14. M Guscin, *The History of the Sudarium of Oviedo*, manuscript copy, p 120.
15. G Heras, J-D Villalain and J-M Rodriguez, *Comparative Study of the Sudarium of Oviedo and the Shroud of Turin*, translated by Mark Guscin, a paper presented at the 3rd International Congress for the Study of the Shroud of Turin, June 1998.
16. Ibid, Footnote 2.
17. *Report on the Second International Conference on the Sudarium of Oviedo, Oviedo, Spain, 13 – 15 April 2007*, published in the BSTS Shroud Newsletter no 65, June 2007.
18. G Heras, J-D Villalain and J-M Rodriguez, *Comparative Study of the Sudarium of Oviedo and the Shroud of Turin*, translated by Mark Guscin, a paper presented at the 3rd International Congress for the Study of the Shroud of Turin, June 1998.
19. C Barta, *The Sudarium of Oviedo and the Man on the Shroud's Ponytail*, a paper published in the BSTS Shroud Newsletter issue #66, December 2007.
20. M Guscin, *The History of the Sudarium of Oviedo*, manuscript copy, p 118. Also on p 103, where the phrase "there is no test for Christness" is attributed to the late Prof. Alan Adler.
21. *Report on the Second International Conference on the Sudarium of Oviedo, Oviedo, Spain, 13 – 15 April 2007*, published in the BSTS Shroud Newsletter issue #65, June 2007.

Chapter 14

1. M Antonacci, *The Resurrection of the Shroud*, (New York 2000), pp 4 – 5.
2. Dr F T Zugibe, *The Crucifixion of Jesus, a Forensic Enquiry*, (New York 2005), pp 246 – 247.
3. Ibid, p 248.
4. I Wilson, *The Turin Shroud*, (London 1979), p 40.
5. Dr F T Zugibe, *The Crucifixion of Jesus, a Forensic Enquiry*, (New York 2005), p 248.

6. Dr P Barbet, *A Doctor at Calvary*, tr. The Earl of Wicklow, (New York 1953), p 9.
7. I Wilson, *The Blood and the Shroud*, (London 1999), p 184 and Plate 39 b.
8. Dr F T Zugibe, *The Crucifixion of Jesus, a Forensic Enquiry*, (New York 2005), p 251. See also M Antonacci, *The Resurrection of the Shroud*, (New York 2000), pp 68 – 69.
9. I Wilson, *The Blood and the Shroud*, (London 1999), pp 362 – 363.
10. I Wilson, *The Turin Shroud*, (London 1979), pp 73 - 74.
11. I Wilson, *The Blood and the Shroud*, (London 1999), p 362.
12. I Wilson, *The Turin Shroud*, (London 1979), pp 73 - 75.
13. I Wilson, *The Blood and the Shroud*, (London 1999), p 365.
14. R N Rogers, *Studies on the Radiocarbon Sample from the Shroud of Turin*, Thermochimica Acta 425 (2005) p 189.
15. K E Stevenson and G R Habermas, *Verdict on the Shroud*, (Pennsylvania 1982), pp 78 – 79.
16. Ibid, p 76.
17. Fr W Bulst, S.J., *The Pollen Grains on the Shroud of Turin*, Shroud Spectrum International no 10, March 1984. See also M Antonacci, *The Resurrection of the Shroud*, (New York 2000), p 109.
18. L Picknett and C Prince, *The Turin Shroud: In Whose Image?* (London 1994), p 13 and note 31 to Chapter 1.
19. Ibid, p 37.
20. M Antonacci, *The Resurrection of the Shroud*, (New York 2000), p 109.
21. E Jumper, K Stevenson Jr and J Jackson, *Images of Coins on a Burial Cloth?*, article in The Numismatist, July 1978.
22. Ibid.
23. K E Stevenson and G R Habermas, *Verdict on the Shroud*, (Pennsylvania 1982), p 77.
24. Ibid, pp 90 – 92.
25. Ibid, p 83.

Chapter 15

1. J H Heller, *Shroud of Mystery*, a condensation of *Report on the Shroud of Turin*, published in the Readers' Digest, January 1984, pp 173 - 179.
2. K F Weaver, *The Mystery of the Shroud*, an article in National Geographic, vol 157, no 6, June 1980, p 749.
3. J H Heller, *Shroud of Mystery*, a condensation of *Report on the Shroud of Turin*, published in the Readers' Digest, January 1984, pp 179 - 180.
4. The members of the STURP team are listed at *www.shroud.com/78team.htm*.
5. J H Heller, *Shroud of Mystery*, a condensation of *Report on the Shroud of Turin*, published in the Readers' Digest, January 1984, p 180.
6. I Wilson, *The Blood and the Shroud*, (London 1999), p 367.
7. J H Heller, *Shroud of Mystery*, a condensation of *Report on the Shroud of Turin*, published in the Readers' Digest, January 1984, pp 180 - 181.
8. I Wilson, *The Blood and the Shroud*, (London 1999), pp 367 – 368.
9. L A Schwalbe and R N Rogers, *Physics and Chemistry of the Shroud of Turin: A Summary of the 1978 Investigation*, published in Analytica Chimica Acta, 135 (1982) 3 – 49, by the Elsevier Scientific Publishing Company, Amsterdam, p 9.
10. K E Stevenson and G R Habermas, *Verdict on the Shroud*, (Pennsylvania 1982), p 99.
11. L A Schwalbe and R N Rogers, *Physics and Chemistry of the Shroud of Turin: A Summary of the 1978 Investigation*, published in Analytica Chimica Acta, 135 (1982) 3 – 49, by the Elsevier Scientific Publishing Company, Amsterdam, pp 10 - 14.
12. J H Heller, *Shroud of Mystery*, a condensation of *Report on the Shroud of Turin*, published in the Readers' Digest, January 1984, p 184.
13. Ibid, p 184.
14. L A Schwalbe and R N Rogers, *Physics and Chemistry of the Shroud of Turin: A Summary of the 1978 Investigation*, published in Analytica Chimica Acta, 135 (1982) 3 – 49, by the

Elsevier Scientific Publishing Company, Amsterdam, pp 23 – 24.
15. Ibid, p 44.
16. Ibid, p 35.
17. K E Stevenson and G R Habermas, *Verdict on the Shroud*, (Pennsylvania 1982), p 119.
18. L A Schwalbe and R N Rogers, *Physics and Chemistry of the Shroud of Turin: A Summary of the 1978 Investigation*, published in Analytica Chimica Acta, 135 (1982) 3 – 49, by the Elsevier Scientific Publishing Company, Amsterdam, pp 25 – 28.
19. J H Heller, *Shroud of Mystery*, a condensation of *Report on the Shroud of Turin*, published in the Readers' Digest, January 1984, p 185.
20. Ibid, p 168.
21. I Wilson, *The Turin Shroud*, (London 1979), pp 263 - 265.
22. I Wilson, *The Blood and the Shroud*, (London 1999), pp 91 - 92.
23. J H Heller, *Shroud of Mystery*, a condensation of *Report on the Shroud of Turin*, published in the Readers' Digest, January 1984, pp 191 - 194.
24. Dr Heller has pointed out that sub-micron particles of iron oxide have been in existence since the dawn of time. In addition, the Shroud had been under lock and key in Turin since the late sixteenth century. The possibility of anybody having had access to it since 1800 to touch it up does not exist. In any event pre-1800 paintings of the Shroud look just like post-1800 paintings and photographs.
25. Ibid, pp 197 – 198.
26. I Wilson, *The Blood and the Shroud*, (London 1999), pp 92 - 93.
27. J H Heller, *Shroud of Mystery*, a condensation of *Report on the Shroud of Turin*, published in the Readers' Digest, January 1984, pp 194 - 196.
28. Ibid, pp 198 – 200.
29. Ibid, pp 199 – 202.
30. K E Stevenson and G R Habermas, *Verdict on the Shroud*, (Pennsylvania 1982), pp 110 – 114. See also L A Schwalbe and R N Rogers, *Physics and Chemistry of the Shroud of Turin:*

A Summary of the 1978 Investigation, published in Analytica Chimica Acta, 135 (1982) 3 – 49, by the Elsevier Scientific Publishing Company, Amsterdam, pp 28 - 30.
31. L A Schwalbe and R N Rogers, *Physics and Chemistry of the Shroud of Turin: A Summary of the 1978 Investigation,* published in Analytica Chimica Acta, 135 (1982) 3 – 49, by the Elsevier Scientific Publishing Company, Amsterdam, pp 12 - 13.
32. Ibid, pp 13 – 16.
33. J H Heller, *Shroud of Mystery,* a condensation of *Report on the Shroud of Turin,* published in the Readers' Digest, January 1984, pp 206 - 207.
34. *A Summary of STURP's Conclusions,* published at *www.shroud.com/78conclu.htm.*
35. Doubt has also been cast recently on Dr McCrone's findings on the Vinland Map. Scientists from the University of California at Davis have found that the map contained only trace amounts of titanium, well over one thousand times less than the amount claimed by Dr McCrone and consistent with amounts that occur in nature. While this by no means authenticates the Vinland Map as a genuine pre-Columbian artifact, it casts yet further doubt on Dr McCrone's scientific methods. See M Antonacci, *The Resurrection of the Shroud,* (New York 2000), Appendix A, pp 259 – 260.

Chapter 16

1. I Wilson, *The Blood and the Shroud,* (London 1999), p 9.
2. Ibid, p 377.
3. Ibid, p 367-8. Also p 207.
4. P E Damon et al, *Radiocarbon Dating of the Shroud of Turin,* a paper reprinted from Nature, vol 337, no 6208, pp 611 – 615, 16 February 1989.
5. I Wilson, *The Blood and the Shroud,* (London 1999), pp 208 - 212.
6. P E Damon et al, *Radiocarbon Dating of the Shroud of Turin,* a paper reprinted from Nature, vol 337, no 6208, pp 611 – 615, 16 February 1989. See also I Wilson, *The Blood and the Shroud,* (London 1999), p 7.

7. P E Damon et al, *Radiocarbon Dating of the Shroud of Turin,* a paper reprinted from Nature, vol 337, no 6208, pp 611 – 615, 16 February 1989.
8. W Meacham, *Radiocarbon Measurement and the Age of the Turin Shroud: Possibilities and Uncertainties,* from the Proceedings of the Symposium "Turin Shroud – Image of Christ?", Hong Kong, March 1986.
9. I Wilson, *The Blood and the Shroud,* (London 1999), p 211.
10. Ibid, p 221.
11. Ibid, pp 263 – 266.
12. Dr L A Garza-Valdes, *The DNA of God?,* (New York 2001), pp 17 - 20.
13. Ibid, pp 24 – 28.
14. Ibid, p 49.
15. Ibid, pp 51 – 52.
16. I Wilson, *The Blood and the Shroud,* (London 1999), p 78.
17. P E Damon et al, *Radiocarbon Dating of the Shroud of Turin,* a paper reprinted from Nature, vol 337, no 6208, pp 611 – 615, 16 February 1989.
18. J G Marino and M S Benford, *Evidence for the Skewing of the C-14 Dating of the Shroud of Turin due to Repairs,* a paper presented at the Congress "Sindone 2000" in August 2000.
19. J G Marino and M S Benford, *Historical Support of a 16th Century Restoration in the Shroud C-14 Sample Area,* a paper published in 2002.
20. R N Rogers and A Arnaldi, *Scientific Method Applied to the Shroud of Turin: A Review,* a paper published in 2002.
21. R N Rogers, *Comments on Benford-Marino Hypothesis,* British Society of the Turin Shroud (BSTS) Newsletter no 54, November 2001, quoted in M S Benford and J G Marino, *Textile Evidence Supports Skewed Radiocarbon Date of Shroud of Turin,* a paper published in 2002. (Issue #54 of the BSTS Newsletter is not publicly available, due to controversy over certain of its contents – personal e-mail from Barrie Schwortz to the author. For this reason I have had to rely on a reference to it in the Benford/Marino paper.)
22. M S Benford and J G Marino, *Textile Evidence Supports Skewed Radiocarbon Date of Shroud of Turin,* a paper published in 2002.

23. R N Rogers and A Arnaldi, *Scientific Method Applied to the Shroud of Turin: A Review*, a paper published in 2002. See Figure 17.
24. J L Brown, *Microscopical Investigation of Selected Raes Threads from the Shroud of Turin*, a paper published in 2005.
25. M S Benford and J G Marino, *New Historical Evidence Explaining the "Invisible Patch" in the 1988 C-14 Sample Area of the Turin Shroud*, a paper published in 2005. See also BSTS Shroud Newsletter no 62, December 2005, p 12.
26. M Flury-Lemberg, *The Invisible Mending of the Shroud in Theory and Reality*, a paper published in the BSTS Shroud Newsletter #65, June 2007.
27. M S Benford and J G Marino, *Invisible mending and the Turin Shroud: Historical and Scientific Evidence*, a paper presented at the Shroud Science Group International Conference held at Ohio State University in August 2008.
28. M S Benford and J G Marino, *Discrepancies in the Radiocarbon Dating Area of the Turin Shroud*, Chemistry Today, Vol 26, no 4, July-August 2008.
29. P Savarino and B Barberis, *Shroud, Carbon Dating and Calculus of Probabilities*, published in 1998 and quoted in M S Benford and J G Marino, *New Historical Evidence Explaining the "Invisible Patch" in the 1988 C-14 Sample Area of the Turin Shroud*, a paper published in 2005.
30. T P Campbell, *Tapestry in the Renaissance: Art and Magnificence*, New Haven and London: Yale University Press, quoted in M S Benford and J G Marino, *New Historical Evidence Explaining the "Invisible Patch" in the 1988 C-14 Sample Area of the Turin Shroud*, a paper published in 2005.
31. I Wilson, *Whatever Happened to Margaret of Austria's 'Shroud relic' Bequest*, article in the June 2000 Newsletter of the British Society for the Turin Shroud, quoted in M S Benford and J G Marino, *New Historical Evidence Explaining the "Invisible Patch" in the 1988 C-14 Sample Area of the Turin Shroud*, a paper published in 2005.
32. Further details of the historical background to these events are given in M S Benford and J G Marino, *New Historical Evidence Explaining the "Invisible Patch" in the 1988 C-14 Sample Area of the Turin Shroud*, a paper published in 2005.

33. R N Rogers and A Arnoldi, *Scientific Method Applied to the Shroud of Turin: A Review*, a paper published in 2002.
34. Ibid.
35. R N Rogers, *Studies on the Radiocarbon Sample from the Shroud of Turin*, published in Thermochimica Acta 425 (2005) pp 189 – 194.

Chapter 17

1. A Summary of STURP's Conclusions. www.shroud.com/78conclu.htm.
2. Ibid.
3. Ibid.
4. E J Jumper, A D Adler, J P Jackson, S F Pellicori, J H Heller and J R Druzik, *A Comprehensive Examination of the Various Stains and Images on the Shroud of Turin*, Archaeological Cemistry III, ACS Advances in Chemistry No 205, J B Lambert, Editor, Chapter 22, American Chemical Society, Washington DC, 1984, pp 447 – 476, referred to in G R Lavoie, *Resurrected*, (Thomas More, Allen, Texas, 2000), pp 63 – 64.
5. A D Adler, *The Nature of the Body Images on the Shroud of Turin*, a paper published in 1999.
6. G R Lavoie, *Resurrected*, (Thomas More, Allen, Texas, 2000), pp 97 - 104.
7. E J Jumper, A D Adler, J P Jackson, S F Pellicori, J H Heller and J R Druzik, *A Comprehensive Examination of the Various Stains and Images on the Shroud of Turin*, Archaeological Cemistry III, ACS Advances in Chemistry No 205, J B Lambert, Editor, Chapter 22, American Chemical Society, Washington DC, 1984, pp 447 – 476, referred to in G R Lavoie, *Resurrected*, (Thomas More, Allen, Texas, 2000), p 62.
8. G R Lavoie, *Resurrected*, (Thomas More, Allen, Texas, 2000), pp 114 - 115.
9. G R Lavoie, B B Lavoie and A D Adler, *Blood on the Shroud of Turin, Part III*, published in Shroud Spectrum International, Sept 1986, referred to in J P Jackson, *Is the Image on the Shroud due to a Process Heretofore Unknown to Modern Science?*, published in Shroud Spectrum International no 34, March 1990.

10. J P Jackson, *Is the Image on the Shroud due to a Process Heretofore Unknown to Modern Science?*, published in Shroud Spectrum International no 34, March 1990.
11. Ibid.
12. A D Acetta, K Lyons and J P Jackson, *Nuclear Medicine and its Relevance to the Shroud of Turin*, an undated paper.
13. G F Carter, *Formation of the Image on the Shroud of Turin by X-Rays: A New Hypothesis*, ACS Advances in Chemistry No 205, Archaeological Chemistry III, edited by J B Lambert, American Chemical Society, 1984, pp 425 – 446, quoted in M Antonacci, *The Resurrection of the Shroud*, (New York 2000), p 213.
14. M Antonacci, *The Resurrection of the Shroud*, (New York 2000), p 214.
15. Ibid, pp 216 – 218.
16. R N Rogers, *Comments on the Book "The Resurrection of the Shroud" by Mark Antonacci*, a paper published in 2001.
17. M Antonacci, *The Resurrection of the Shroud*, (New York 2000), pp 222 – 232.
18. R N Rogers, *Comments on the Book "The Resurrection of the Shroud" by Mark Antonacci*, a paper published in 2001.
19. M Antonacci, *The Resurrection of the Shroud*, (New York 2000). On page 222 he quotes Dr Kitty Little, a retired nuclear physicist as referring to "something in the way of a nuclear disintegration, acting almost instantaneously, as with the flash from a nuclear explosion" as being the possible source of ionizing radiation in the Shroud image formation process. He also quotes Prof Wesley McDonald as "summarizing the viewpoints of many scientists" in describing the burst of energy from the body being like "a pulsed laser beam caused by dematerialization of the body into energy in a millisecond."
20. R N Rogers, *Comments on the Book "The Resurrection of the Shroud" by Mark Antonacci*, a paper published in 2001.
21. M Antonacci, *The Resurrection of the Shroud*, (New York 2000), Note 62 to Chapter 10, pp 310 – 311.
22. R N Rogers, *Comments on the Book "The Resurrection of the Shroud" by Mark Antonacci*, a paper published in 2001.

23. G Fanti and R Maggiolo, *The Double Superficiality of the Frontal Image of the Turin Shroud,* Journal of Optics A: Pure and Applied Optics 6 (2004), pp 491 – 503.
24. M Latendresse, *The Turin Shroud was not Flattened Before the Images Formed and no Major Image Distortions Necessarily Occur from a Real Body,* a paper presented at the Third International Dallas Conference on the Shroud of Turin, September 2005.
25. G Fanti, F Lattarulo and O Scheuermann, *Body Image Formation Hypotheses based on Corona Discharge,* a paper presented at the Third International Dallas Conference on the Shroud of Turin, September 2005.
26. Ibid.
27. G R Lavoie, *A Forensic Medical Study of the Shroud of Turin that will Present New Information to Further Illustrate the Complexity of Image Formation,* a paper presented at the Third International Dallas Conference on the Shroud of Turin, September 2005.
28. G R Lavoie, *Resurrected,* (Thomas More, Allen, Texas, 2000), pp 41. Figure 2, on page 40, showing the back image, positive image, backlighted, appears to bear this out. The buttocks and calf muscles of the image appear firm and rounded.
29. Ibid, pp 134 – 135. Figures 9 and 12 on page 134 illustrate Dr Lavoie's argument.
30. Ibid, Chapter 8, pp 127 – 138.
31. Dr F T Zugibe, *The Crucifixion of Jesus, a Forensic Enquiry,* (New York 2005) p 189.
32. Ibid, p 212.
33. G R Lavoie, *Resurrected,* (Thomas More, Allen, Texas, 2000), p 145.
34. I Wilson, *The Blood and the Shroud,* (London 1999), pp 245 - 251. See also M Antonacci, *The Resurrection of the Shroud,* (New York 2000), pp 84 – 93.
35. L Picknett and C Prince, *Turin Shroud – in whose Image?* (London 1994). See also I Wilson, *The Blood and the Shroud,* (London 1999), pp 242 – 244, and M Antonacci, *The Resurrection of the Shroud,* (New York 2000), pp 84 – 93.
36. T Ware, *The Orthodox Church,* (Penguin Books, 1963), p 75.

Chapter 18

1. M Antonacci, *The Resurrection of the Shroud*, (New York 2000), pp 73 - 76.
2. Ibid, pp 76 – 77.
3. *The Shroud of Moscow, Idaho,* an article by Joan Opyr carried in New West Living (the Voice of the Rocky Mountains) on 22 March 2005, available at *www.newwest.net/index.php/topic/article/621/C84/L40.*
4. M Antonacci, *The Resurrection of the Shroud*, (New York 2000), pp 77 - 79.
5. Dr L A Garza-Valdes, *The DNA of God?*, (New York 2001), pp 55 - 59.
6. R N Rogers and A Arnoldi, *Scientific Method Applied to the Shroud of Turin: A Review,* a paper published in 2002.
7. L A Schwalbe and R N Rogers, *Physics and Chemistry of the Shroud of Turin: A Summary of the 1978 Investigation,* a paper published in Analytica Chemica Acta, 135 (1982) 3 – 49, published by the Elsevier Scientific Publishing Company, Amsterdam.
8. R N Rogers and A Arnoldi, *Scientific Method Applied to the Shroud of Turin: A Review,* a paper published in 2002.
9. Ibid.
10. Ibid.
11. Ibid.
12. Ibid.
13. Ibid.
14. R N Rogers, *The Shroud of Turin: Radiation Effects, Ageing and Image Formation,* a paper published in 2005.
15. J C Iannone, *The Mystery of the Shroud of Turin,* (New York 1998), p 25. Oswald Scheuermann later collaborated with Prof Giulio Fanti on studies on the possibility of the image on the Shroud being formed by a corona discharge process.
16. J C Iannone, *Floral Images and Pollen Grains on the Shroud of Turin: An Interview with Dr Alan Whanger and Dr Avinoam Danin,* published in 1999.
17. A Danin, *The Origin of the Shroud of Turin from the Near East as Evidenced by Plant Images and by Pollen Grains,* a paper published in 1998.

18. Ibid. See also A Danin, *Pressed Flowers: Where Did the Shroud of Turin Originate? A Botanical Quest*, a paper published in ERETZ Magazine, November/December 1997.
19. J C Iannone, *The Mystery of the Shroud of Turin*, (New York 1998), pp 29 – 30.
20. R N Rogers, *Comments on the Book "The Resurrection of the Shroud" by Mark Antonacci*, a paper published in 2001.

Chapter 19

1. W Meacham, *The Rape of the Turin Shroud*, (2005), p 150.
2. Report in *La Voce del Popolo*, Sunday 13 September 1992, quoted in W Meacham, *The Rape of the Turin Shroud*, (2005), p 151. See also *1992 – 2002: Ten Years of Important Events in the History of the Holy Shroud*, a paper by B Barberis and P Savarino presented at the Third International Dallas Conference on the Shroud of Turin, September 2005.
3. B Barberis and P Savarino, *1992 – 2002: Ten Years of Important Events in the History of the Holy Shroud*, a paper presented at the Third International Dallas Conference on the Shroud of Turin, September 2005.
4. W Meacham, *The Rape of the Turin Shroud*, (2005), pp 153 – 155.
5. B Barberis and P Savarino, *1992 – 2002: Ten Years of Important Events in the History of the Holy Shroud*, a paper presented at the Third International Dallas Conference on the Shroud of Turin, September 2005. See also W Meacham, *The Rape of the Turin Shroud*, (2005), pp 162 – 164.
6. B Schwortz, *An Editorial Comment by Barrie Schwortz on an article published by Zenit News Service on 10 May 2001, entitled "New Study Backs Authenticity of Turin Shroud – Scanner Captures Image of Reverse Side"*. This comment was published on his 2001 Website News Page at *www.shroud.com/late01.htm*.
7. W Meacham, *The Rape of the Turin Shroud*, (2005), p 167.
8. P C Maloney, *The New Restoration of the Shroud: Conservation, History or Science?*, a paper published in 2003 at *www.shroud.com/restored.htm*.
9. BSTS Shroud Newsletter #62, December 2005, pp 7 – 8.

10. P C Maloney, *The New Restoration of the Shroud: Conservation, History or Science?*, a paper published in 2003 at *www.shroud.com/restored.htm*.
11. e-mail from Barrie Schwortz to the author dated 20 November 2007.
12. W Meacham, *The Rape of the Turin Shroud*, (2005), pp 153 - 155.
13. B Schwortz, *An Editorial Comment by Barrie Schwortz on an article published by Zenit News Service on 10 May 2001, entitled "New Study Backs Authenticity of Turin Shroud – Scanner Captures Image of Reverse Side"*. See also Rev A Dreisbach, Jr, *Turin: Then and Now*, a paper published in 2003 at *www.shroud.com/restored.htm*.
14. J C Cruz, *Relics* (Huntingdon, Indiana, 1984), p 6.
15. Quoted in W Meacham, *The Rape of the Turin Shroud*, (2005), pp 104 – 105.
16. It is unfortunate that much of Prof Meacham's criticism has been expressed in terms that can only be described as intemperate and, not infrequently, insulting to those with whom he disagrees. His book, *The Rape of the Turin Shroud*, (2005), describes events both leading up to the restoration and subsequent to it and is of substantial interest. His language in the book, however, is caustic in the extreme, particularly when making references to researchers who expressed approval of the work done by the restoration. He describes one researcher from France, during a press conference following the presentation of the restored Shroud, as being "engaged in mere sycophancy and (knowing) nothing about real conservation." (Page 193). He goes on to describe Ian Wilson as giving "an even more nauseatingly sweet speech, gushing with compliments". (Page 193). Ian Wilson, in a personal e-mail to me dated 18 November 2007, referred to Meacham's "hysteria" and described him as "ultra-aggressive in his attitude towards Turin". He feels that Prof Meacham's actions and attitude have been counter-productive in that they have made the Church authorities in Turin and Rome less inclined to co-operate in further international scientific research on the Shroud. Barrie Schwortz, on the other hand, is of the view that Prof Meacham is "completely justified in most of his criticism"

(also in a personal e-mail to me, dated 20 November 2007) but he does concede that "he gets personal sometimes, and I feel that is a mistake".
17. These various quotations are taken from a book about the restoration written by Mme Flury-Lemburg which was published by the Archdiocese of Turin in July 2003. They are quoted in W Meacham, *The Rape of the Turin Shroud*, (2005), pp 202 - 203. I have not had access to a copy of Mme Flury-Lemburg's book and am reliant on second-hand quotations from it.
18. I have to observe that I am extremely sceptical of the ability of any person to observe the image of flowers on the Shroud itself from a distance, using binoculars, let alone identify them. I would have thought that such observations, if indeed they are possible, could only be made by detailed close-up staudy.
19. W Meacham, *The Rape of the Turin Shroud*, (2005), pp 203 - 208. Prof Meacham's also quotes Dr Ray Rogers in support of his views. He cites Dr Rogers as being "unequivocal that there is **absolutely** no threat (to the Shroud) from the charred material or carbon dust in any chemical sense…." (p 207).
20. Ibid, pp 208 – 209.
21. P C Maloney, *The New Restoration of the Shroud: Conservation, History or Science?*, a paper published in 2003 at *www.shroud.com/restored.htm*.
22. W Meacham, *The Rape of the Turin Shroud*, (2005), p 217.
23. P C Maloney, *The New Restoration of the Shroud: Conservation, History or Science?*, a paper published in 2003 at *www.shroud.com/restored.htm*.
24. e-mail from Barrie Schwortz to the author dated 20 November 2007.
25. R N Rogers, *Frequently Asked Questions (FAQs)*, a paper published in 2004, section 15.
26. B Schwortz, *2002 Late Breaking Website News Page*, published at *www.shroud.com*.
27. BSTS Shroud Newsletter #62, December 2005, pp 3 - 17.
28. W Meacham, *The Rape of the Turin Shroud*, (2005), p 231.
29. Ibid, pp 249 – 250.
30. D Scavone, *Comments on the Intervention Done on the Shroud of Turin in June-July 2002*, a paper published in 2003 at *www.shroud.com/restored.htm*.

31. S E Jones, *A Proposal to Radio Carbon Date the Pollen on the Shroud of Turin,* a paper published in the BSTS Shroud Newsletter #66, December 2007.
32. G Fanti, *Comments on the "Restoration of the Shroud" Done in June-July 2002,* a paper published in 2003 at *www.shroud.com/restored.htm.*